FE MECHANICAL EXAMS

TWO PRACTICE EXAMS WITH STEP-BY-STEP SOLUTIONS

MOHAMMAD IQBAL, D.Sc., PE, SE, Esq.
ALI A. IQBAL, PE

Report Errors for This Book

PPI is grateful to every reader who notifies us of a possible error. Your feedback allows us to improve the quality and accuracy of our products. Report errata at **ppi2pass.com**.

FE MECHANICAL EXAMS:
TWO PRACTICE EXAMS WITH STEP-BY-STEP SOLUTIONS

Current release of this edition: 1

Release History

date	edition number	revision number	update
Sept 2022	1	1	New edition.

Printed in the United States of America.

PPI
ppi2pass.com

ISBN: 978-1-59126-876-5

Table of Contents

Preface and Acknowledgments

Our objective in writing *FE Mechanical Exams: Two Practice Exams* was to create a book that (1) provides exam-like practice for the NCEES FE Mechanical exam and (2) familiarizes you with the *NCEES FE Reference Handbook* (*Handbook*), the only reference you are allowed to use during the exam. We believe this book accomplishes both objectives.

After we published *FE Civil Exams: 5 Practice Exams*, we determined that a similar book was needed to help readers preparing for the NCEEES FE Mechanical Engineering (ME) examination. For this new book, we have created 220 questions covering two full-length practice exams. The questions have the same difficulty and format as the FE Mechanical Engineering exam. The FE Mechanical Engineering exam has evolved in its format and content, so we've included alternative item type (AIT) problems in the book. Each exam has 110 problems.

In order to better simulate the exam experience, the questions are limited to the 14 knowledge areas and their subknowledge areas listed by NCEES in its specifications for the exam. The number of questions in each knowledge area follows the average weight that NCEES has given to the knowledge/subknowledge area in its exam specifications. The equations used in the solutions have the same notations as given in the *Handbook*. Detailed tables in the introductory portion of this book will help you learn where to look in the *Handbook* to find the data and equations you'll need during the exam.

This book represents a great deal of effort from a great many people. We would like to acknowledge the team at PPI, including

Editorial: Bilal Baqai, Susan Bedell, Scott Marley, Pooja Thakur, Grace Wong, Michael Wordelman

Art and Cover Design: Tom Bergstrom

Production: Sean Woznicki, Richard Iriye, Beth Christmas, Kim Wimpsett, Kim Burton-Weisman

Project Management: Jeri Jump

Content and Product: Anna Howland; Joseph Konczynski, PE; Nicole Evans

Publishing Systems: Sam Webster

We'd also like to thank our technical reviewers and calculation checkers: S. Sandler, PE; Lise Palmer, PE

We hope this book helps you with your exam preparation. Though we made every effort to ensure the technical accuracy of our problems, any mistakes you find are ours alone. Please submit suspected errors using PPI's errata reporting website, **ppi2pass.com/errata**.

Mohammad Iqbal, D.Sc., PE, SE, Esq.
Ali A. Iqbal, PE

How to Use This Book

Use this book to practice solving problems by using the *NCEES FE Reference Handbook* as your sole reference. The *Handbook* is the only reference you may use during the NCEES FE Mechanical exam.

You should be familiar with the *Handbook* before you begin to take the practice exams in this book. Know which exam knowledge areas cover which particular mechanical engineering concepts and formulas. Then use the review materials of your choice to study those concepts and formulas.

Attempt to solve each problem on your own, using only the *Handbook* as a reference. If you are unsure where to begin on any particular problem, use the tables found in the "*NCEES Handbook* Sections by Problem Number" section of this book to find the sections in the *Handbook* that each problem references. If you are still unable to solve the problem, review the first few lines of the solution, and see if you can do the rest on your own. If you cannot solve several similar problems, go back and review your study materials, and then retry the problems. Once you feel comfortable solving problems within a given exam, move on to the next exam in the book.

Give yourself plenty of time to study and test yourself; begin at least three months before the exam. The time required to study will vary based on how long you have been out of school and whether your knowledge is general or specialized. Create a study schedule for yourself and stick to it. If you give yourself adequate time to study, if you study hard, and if you work the problems until you are sure you understand how to navigate the *Handbook* quickly and efficiently, you will give yourself the best possible chance to pass the NCEES FE Mechanical exam.

NCEES Handbook Sections by Problem Number

This section shows the relationship between problems in this book and the *NCEES FE Reference Handbook*, version 10.2, published for the FE Mechanical exam. The *Handbook* is the only material allowed as a reference during the exam. The exact titles of chapters and sections from the *Handbook* are given in this section so that they can be used as search terms to find the equations, tables, and other information that you need. If you use these search terms as you study, you will come to learn the ones you use most often, and this will save you time during the exam.

When taking these practice exams, use the *Handbook* as you would for the real exam. Then, while reviewing the solutions, use the tables that follow to find the relevant chapters and sections of the *Handbook*. Be sure to especially familiarize yourself with any problems that you missed.

A word of caution: Subject matter that does *not* appear in the *Handbook* may still appear on the exam. In the table shown in this section, a *Handbook* chapter without a *Handbook* section indicates that the subject matter as a whole is important for the exam but that there is no specific *Handbook* equation or table that you need for that problem. Familiarize yourself with the exam specifications, in addition to the *Handbook*, to prepare for the sorts of questions you may see.

NCEES Handbook Sections by Problem Number: Exam 1

problem	chapter	section
1	Mathematics	Straight Line
2	Mathematics	Indefinite Integrals
3	Mathematics	First-Order Linear Homogeneous Differential Equations with Constant Coefficients
4	Mathematics	Vectors
5	Mathematics	Taylor's Series
6	Mathematics	
7	Mathematics	Determinants
8	Mathematics	L'Hospital's Rule (L'Hôpital's Rule)
9	Engineering Probability and Statistics	Permutations and Combinations
10	Engineering Probability and Statistics	Dispersion, Mean, Median, and Mode Values
11	Engineering Probability and Statistics	Dispersion, Mean, Median, and Mode Values
12	Engineering Probability and Statistics	Linear Regression and Goodness of Fit
13	Engineering Probability and Statistics	Laws of Probability
14	Ethics and Professional Practice	Code of Ethics
15	Ethics and Professional Practice	Model Rules
16	Ethics and Professional Practice	Intellectual Property
17	Ethics and Professional Practice	Societal Considerations
18	Ethics and Professional Practice	Societal Considerations
19	Ethics and Professional Practice	Model Rules
20	Engineering Economics	Single Payment Compound Amount
21	Engineering Economics	Interest Rate Tables
22	Engineering Economics	Differential Calculus
23	Engineering Economics	Breakeven Analysis
24	Engineering Economics	Depreciation
25	Engineering Economics	Interest Rate Tables
26	Electrical and Computer Engineering	Kirchhoff's Laws
27	Electrical and Computer Engineering	Resistors in Series and Parallel
28	Electrical and Computer Engineering	Kirchhoff's Laws
29	Electrical and Computer Engineering	Electrostatic Fields
30	Electrical and Computer Engineering	AC Machines
31	Statics	Resolution of a Force
32	Statics	Moments (Couples)
33	Statics	Equilibrium Requirements
34	Statics	Plane Truss: Method of Joints
35	Civil Engineering	Stability, Determinacy, and Classification of Structures
36	Statics	Centroids of Masses, Areas, Lengths, and Volumes
37	Statics	Moment of Inertia

NCEES Handbook Sections by Problem Number: Exam 1 (cont'd)

problem	chapter	section
38	Statics	Belt Friction
39	Statics	Friction
40	Statics	Equilibrium Requirements
41	Dynamics	Constant Acceleration
42	Dynamics	Friction
43	Dynamics	Kinetic Energy
44	Dynamics	Impulse and Momentum Impact
45	Dynamics	Free and Forced Vibration
46	Dynamics	Linear Momentum
47	Dynamics	Kinetic Energy
48	Dynamics	Constant Acceleration Friction
49	Dynamics	Friction
50	Dynamics	Kinetics of a Rigid Body
51	Mechanics of Materials	Uniaxial Loading and Deformation Material Properties
52	Mechanics of Materials	Shear Stress-Strain
53	Mechanics of Materials	Columns
	Statics	Area Moment of Inertia
54	Mechanics of Materials	Thermal Deformations
55	Mechanics of Materials	Beams Simply Supported Beam Slopes and Deflections
56	Mechanics of Materials	Torsional Strain
57	Mechanics of Materials	Deflection of Beams
58	Mechanics of Materials	Uniaxial Loading and Deformation
59	Mechanics of Materials	Stresses in Beams
60	Mechanics of Materials	Mohr's Circle—Stress, 2D
61	Mechanics of Materials	Mohr's Circle—Stress, 2D
62	Materials Science/Structure of Matter	Corrosion
63	Materials Science/Structure of Matter	Mechanical
64	Mechanical Engineering	Variable Loading Failure Theories
65	Mechanics of Materials	Uniaxial Stress-Strain
66	Materials Science/Structure of Matter	Composite Materials
67	Materials Science/Structure of Matter	Iron-Iron Carbide Phase Diagram
68	Materials Science/Structure of Matter	Material Properties
69	Fluid Mechanics	Reynolds Number
70	Fluid Mechanics	The Pressure Field in a Static Liquid
71	Dynamics	Particle Kinetics
	Fluid Mechanics	Deflectors and Blades The Impulse-Momentum Principle
72	Fluid Mechanics	Energy Equation
73	Fluid Mechanics	Drag Force Drag Coefficient for Spheres, Disks, and Cylinders
74	Fluid Mechanics	Compressible Flow

NCEES Handbook **Sections by Problem Number: Exam 1 (cont'd)**

problem	chapter	section
75	Mechanics of Materials	Torsional Strain
	Fluid Mechanics	Stress, Pressure, and Viscosity
76	Fluid Mechanics	Consequences of Fluid Flow Principles of One-Dimensional Fluid Flow: Energy Equation
77	Fluid Mechanics	Performance of Components
78	Fluid Mechanics	Deflectors and Blades
79	Thermodynamics	Properties for Two-Phase (vapor-liquid) Systems
80	Thermodynamics	First Law of Thermodynamics
81	Thermodynamics	First Law of Thermodynamics
82	Thermodynamics	Mollier (h, s) Diagram for Steam
83	Mechanical Engineering	Cycles and Processes
84	Thermodynamics	Basic Cycles
85	Thermodynamics	Cycles and Processes Pressure Versus Enthalpy Curves for Refrigerant 410A (USCS units) Refrigerant 410A [R-32/125 (50/50)] Properties
86	Thermodynamics	Entropy
87	Thermodynamics	ASHRAE Psychrometric Chart No. 1
88	Thermodynamics	Heats of Reaction: Combustion Processes
89	Heat Transfer	Basic Heat-Transfer Rate Equations
90	Heat Transfer	Conduction
91	Heat Transfer	Convection
92	Heat Transfer	Radiation
93	Heat Transfer	Transient Conduction Using the Lumped Capacitance Model
94	Heat Transfer	Basic Heat-Transfer Rate Equations
95	Heat Transfer	Conduction Through a Cylindrical Wall
96	Instrumentation, Measurement, and Control	Measurement
97	Instrumentation, Measurement, and Control	Control Systems
98	Dynamics	Free and Forced Vibration
99	Instrumentation, Measurement, and Control	Measurement Uncertainty
100	Instrumentation, Measurement, and Control	Control Systems
101	Mechanical Engineering	Gearing
102	Mechanical Engineering	Modified Goodman Theory
103	Mechanics of Materials	Simply Supported Beam Slopes and Deflections
	Mechanical Engineering	Springs
104	Mechanical Engineering	Mechanical Springs
105	Mechanics of Materials	Cylindrical Pressure Vessel
106	Mechanical Engineering	Bearings
107	Statics	Screw Thread
108	Statics	Belt Friction
109	Mechanics of Materials	Joining Methods
110	Mechanical Engineering	Manufacturability

NCEES Handbook Sections by Problem Number: Exam 2

problem	chapter	section
111	Mathematics	Straight Line
112	Mathematics	Derivatives
113	Mathematics	Second-Order Linear Homogeneous Differential Equations with Constant Coefficients
114	Mathematics	Matrices
115	Mathematics	Newton's Method for Root Extraction
116	Mathematics	Software Engineering
117	Mathematics	Test for a Point of Inflection
118	Mathematics	Indefinite Integrals
119	Engineering Probability and Statistics	Permutations and Combinations
120	Engineering Probability and Statistics	Dispersion, Mean, Median, and Mode Values
121	Engineering Probability and Statistics	Dispersion, Mean, Median, and Mode Values
122	Engineering Probability and Statistics	Linear Regression and Goodness of Fit
123	Engineering Probability and Statistics	Test Statistics Normal Distribution (Gaussian Distribution)
124	Ethics and Professional Practice	Code of Ethics
125	Ethics and Professional Practice	Intellectual Property
126	Ethics and Professional Practice	Intellectual Property
127	Ethics and Professional Practice	*Model Rules*, Section 240.15 Rules of Professional Conduct
128	Ethics and Professional Practice	*Model Rules*, Section 240.15 Rules of Professional Conduct Societal Considerations
129	Ethics and Professional Practice	*Model Rules*, Section 240.15 Rules of Professional Conduct
130	Engineering Economics	Capitalized Costs
131	Engineering Economics	Interest Rate Tables
132	Engineering Economics	Breakeven Analysis
133	Engineering Economics	Interest Rate Tables
134	Engineering Economics	Interest Rate Tables
135	Engineering Economics	Interest Rate Tables
136	Electrical and Computer Engineering	Capacitors and Inductors
137	Electrical and Computer Engineering	Capacitors and Inductors
138	Electrical and Computer Engineering	Source Equivalents
139	Electrical and Computer Engineering	Effective or RMS Values
140	Electrical and Computer Engineering	DC Machines Rotating Machines (General)
141	Statics	Equilibrium Requirements
142	Statics	Resolution of a Force
143	Statics	Systems of Forces
144	Statics	Statically Determinate Truss
145	Statics	Plane Truss: Method of Sections
146	Statics	Centroids of Masses, Areas, Lengths, and Volumes
147	Statics	Area Moment of Inertia
148	Statics	Friction
149	Statics	Belt Friction Rotating Machines (General)
150	Statics	Plane Truss: Method of Joints
151	Dynamics	Work
152	Dynamics	Particle Kinetics

NCEES Handbook Sections by Problem Number: Exam 2 (cont'd)

problem	chapter	section
153	Dynamics	Particle Kinetics
154	Dynamics	Potential Energy
155	Dynamics	Constant Acceleration
156	Dynamics	Kinematics of a Rigid Body
157	Statics	Centroids of Masses, Areas, Lengths, and Volumes
	Dynamics	Kinematics of a Rigid Body Mass Moment of Inertia
158	Materials Science/Structure of Matter	Impact Test
	Dynamics	Potential Energy
159	Dynamics	Kinetic Energy and Work
160	Dynamics	Free and Forced Vibration
161	Statics	Area Moment of Inertia
	Mechanics of Materials	Beams: Shearing Force and Bending Energy
162	Mechanics of Materials	Mohr's Circle—Stress, 2D
163	Mechanics of Materials	Uniaxial Loading and Deformation Material Properties
164	Mechanics of Materials	Deflection of Beams
165	Mechanics of Materials	Torsional Strain
166	Mechanics of Materials	Shear Stress-Strain
167	Mechanics of Materials	Thermal Deformations
168	Mechanical Engineering	Fastener Groups in Shear
169	Mechanics of Materials	Simply Supported Beam Slopes and Deflections
170	Mechanics of Materials	Columns
	Civil Engineering	W Shapes Dimensions and Properties
171	Civil Engineering	Beam Stiffness and Moment Carryover
172	Materials Science/Structure of Matter	Relationship Between Hardness and Tensile Strength
173	Materials Science/Structure of Matter	Properties of Materials: Mechanical
174	Materials Science/Structure of Matter	Representative Values of Fracture Toughness
175	Materials Science/Structure of Matter	Hardenability
176	Materials Science/Structure of Matter	Thermal and Mechanical Processing
177	Chemistry and Biology	Standard Oxidation Potentials for Corrosion Reactions
	Materials Science/Structure of Matter	Corrosion
178	Mechanics of Materials	Average Mechanical Properties of Typical Engineering Materials
179	Units and Conversion Factors	
180	Fluid Mechanics	Forces on Submerged Surfaces and the Center of Pressure
181	Fluid Mechanics	Deflectors and Blades Properties of Water
182	Fluid Mechanics	Stress, Pressure, and Viscosity
183	Fluid Mechanics	Drag Force
184	Thermodynamics	
	Fluid Mechanics	Isentropic Flow Relationships
185	Fluid Mechanics	Consequences of Fluid Flow
186	Fluid Mechanics	Fluid Flow Machinery
187	Fluid Mechanics	Performance of Components
188	Fluid Mechanics	Dimensional Homogeneity
189	Thermodynamics	*PVT* Behavior
190	Thermodynamics	Closed Thermodynamic System

NCEES Handbook Sections by Problem Number: Exam 2 (cont'd)

problem	chapter	section
191	Thermodynamics	Ideal Gas Thermal and Physical Property Tables
192	Thermodynamics	Steady-Flow Systems
193	Thermodynamics	Mollier (h, s) Diagram for Steam
194	Thermodynamics	Thermal and Physical Property Tables
	Mechanical Engineering	Cycles and Processes
195	Thermodynamics	Basic Cycles P-h Diagram for Refrigerant HFC-134a Common Thermodynamic Cycles
196	Thermodynamics	Special Cases of Steady-Flow Energy Equation
197	Thermodynamics	Psychrometric Chart
198	Chemistry and Biology	
	Thermodynamics	Heats of Reaction
199	Heat Transfer	Thermal Resistance (R) Composite Plane Wall
200	Heat Transfer	Shape Factor (View Factor, Configuration Factor) Relations
201	Heat Transfer	Net Energy Exchange by Radiation between Two Bodies
202	Heat Transfer	Basic Heat-Transfer Rate Equations
203	Materials Science/Structure of Matter	Properties of Metals
	Heat Transfer	Approximate Solution for Solid with Sudden Convection
204	Heat Transfer	Heat Exchangers
205	Heat Transfer	Biot Number
206	Instrumentation, Measurement, and Control	Control Systems
207	Instrumentation, Measurement, and Control	Control Systems
208	Instrumentation, Measurement, and Control	Measurement Uncertainty
209	Instrumentation, Measurement, and Control	Strain Transducers
210	Instrumentation, Measurement, and Control	Temperature Sensors
211	Mechanical Engineering	Power Transmission: Shafts and Axles
212	Mechanical Engineering	Power Transmission: Shafts and Axles
213	Mechanical Engineering	Joining Methods
214	Mechanical Engineering	Springs
215	Statics	
	Mechanical Engineering	Screw Thread
216	Mechanical Engineering	Brake Power
217	Mechanical Engineering	Joining Methods
218	Mechanical Engineering	Industrial and Systems Engineering: Reliability
219	Mechanical Engineering	Geometric Dimensioning and Tolerancing (GD&T)
220	Mechanical Engineering	Geometric Dimensioning and Tolerancing (GD&T)

Exam 1 Instructions

In accordance with the rules established by your state, you may use any approved battery- or solar-powered, silent calculator to work this examination. However, no blank papers, writing tablets, unbound scratch paper, or loose notes are permitted. Sufficient paper will be provided. The *NCEES FE Reference Handbook* is the only reference you are allowed to use during this exam.

You are not permitted to share or exchange materials with other examinees.

You will have six hours in which to work this session of the examination. Your score will be determined by the number of questions that you answer correctly. There is a total of 110 questions. All 110 questions must be worked correctly in order to receive full credit on the exam. There are no optional questions. Each question is worth 1 point. The maximum possible score for this section of the examination is 110 points.

Partial credit is not available. No credit will be given for methodology, assumptions, or work written on scratch paper.

Record all your answers on the Answer Sheet. If you change your answer, make sure your final answer is clearly indicated. Select only one answer per question unless instructed otherwise. Some questions will require you to select more than one correct answer, select a point on a graphic, move items to their correct locations in a table or figure, or enter your response in a blank space. All questions have the same point value regardless of type.

If you finish early, check your work and make sure that you have followed all instructions. After checking your answers, you may submit your answers and leave the examination room. Once you leave, you will not be permitted to return to work or change your answers.

When permission has been given by your proctor, you may begin your examination.

Name: ______________________________
Last First Middle Initial

Examinee number: ______________

Examination Booklet number: ______________

Fundamentals of Engineering Examination

Exam 1

Exam 1 Answer Sheet

1.	(A) (B) (C) (D)
2.	(A) (B) (C) (D)
3.	(A) (B) (C) (D)
4.	(A) (B) (C) (D)
5.	(A) (B) (C) (D)
6.	(A) (B) (C) (D)
7.	(A) (B) (C) (D)
8.	(A) (B) (C) (D)
9.	(A) (B) (C) (D)
10.	(A) (B) (C) (D)
11.	(A) (B) (C) (D)
12.	(A) (B) (C) (D)
13.	(A) (B) (C) (D)
14.	(A) (B) (C) (D) (E)
15.	(A) (B) (C) (D)
16.	(A) (B) (C) (D)
17.	(A) (B) (C) (D)
18.	(A) (B) (C) (D)
19.	(A) (B) (C) (D)
20.	(A) (B) (C) (D)
21.	(A) (B) (C) (D)
22.	(A) (B) (C) (D)
23.	(A) (B) (C) (D)
24.	(A) (B) (C) (D)
25.	(A) (B) (C) (D)
26.	(A) (B) (C) (D)
27.	(A) (B) (C) (D)
28.	(A) (B) (C) (D) (E)
29.	(A) (B) (C) (D)
30.	(A) (B) (C) (D)
31.	(A) (B) (C) (D)
32.	(A) (B) (C) (D)
33.	(A) (B) (C) (D)
34.	(A) (B) (C) (D)
35.	(A) (B) (C) (D)
36.	(A) (B) (C) (D)
37.	(A) (B) (C) (D)
38.	(A) (B) (C) (D) (E)
39.	(A) (B) (C) (D)
40.	(A) (B) (C) (D)
41.	(A) (B) (C) (D)
42.	(A) (B) (C) (D)
43.	(A) (B) (C) (D)
44.	(A) (B) (C) (D)
45.	(A) (B) (C) (D)
46.	(A) (B) (C) (D)
47.	(A) (B) (C) (D)
48.	(A) (B) (C) (D)
49.	(A) (B) (C) (D)
50.	(A) (B) (C) (D)
51.	(A) (B) (C) (D)
52.	(A) (B) (C) (D)
53.	(A) (B) (C) (D)
54.	(A) (B) (C) (D)
55.	(A) (B) (C) (D)
56.	(A) (B) (C) (D)
57.	(A) (B) (C) (D)
58.	(A) (B) (C) (D)
59.	(A) (B) (C) (D)
60.	(A) (B) (C) (D)
61.	(A) (B) (C) (D)
62.	(A) (B) (C) (D)
63.	(A) (B) (C) (D)
64.	(A) (B) (C) (D)
65.	(A) (B) (C) (D)
66.	(A) (B) (C) (D)
67.	(A) (B) (C) (D)
68.	(A) (B) (C) (D)
69.	(A) (B) (C) (D)
70.	(A) (B) (C) (D)
71.	(A) (B) (C) (D)
72.	(A) (B) (C) (D)
73.	(A) (B) (C) (D)
74.	(A) (B) (C) (D)
75.	(A) (B) (C) (D)
76.	________________
77.	(A) (B) (C) (D)
78.	(A) (B) (C) (D)
79.	(A) (B) (C) (D)
80.	(A) (B) (C) (D)
81.	________________
82.	________________
83.	________________
84.	________________
85.	________________
86.	________________
87.	________________
88.	________________
89.	(A) (B) (C) (D)
90.	________________
91.	________________
92.	(A) (B) (C) (D)
93.	(A) (B) (C) (D)
94.	(A) (B) (C) (D)
95.	(A) (B) (C) (D)
96.	(A) (B) (C) (D)
97.	(A) (B) (C) (D)
98.	(A) (B) (C) (D)
99.	________________
100.	(A) (B) (C) (D)
101.	(A) (B) (C) (D)
102.	(A) (B) (C) (D)
103.	(A) (B) (C) (D)
104.	(A) (B) (C) (D)
105.	(A) (B) (C) (D)
106.	(A) (B) (C) (D)
107.	(A) (B) (C) (D)
108.	(A) (B) (C) (D)
109.	(A) (B) (C) (D)
110.	(A) (B) (C) (D)

Exam 1

1. The equation of a line passing through points (1, 2) and (5, 0) is

(A) $2y+x=5$

(B) $y+2x=5$

(C) $y+3x=5$

(D) $3y+x=5$

2. Line $y=2x$ is rotated about the y-axis. The volume of the resulting solid between $y=0$ and $y=10$ is most nearly

(A) $50\pi/3$

(B) $100\pi/3$

(C) $150\pi/3$

(D) $250\pi/3$

3. A differential equation, and its boundary conditions are

$$\frac{dy}{dx}+3y=0;\ y(0)=1$$

The general solution to the equation is

(A) e^{-3t}

(B) e^{3t}

(C) $3e^{t}$

(D) e^{i3t}

4. The area of a parallelogram bounded by vectors from the origin to points (0, 5) and (4, 5) is

(A) 12.5

(B) 16.0

(C) 20.0

(D) 25.0

5. The following series equation is used to approximate the value of sin x, where x is an angle.

$$\sin x \approx x-\frac{x^3}{3!}+\frac{x^5}{5!}+\cdots$$

The value of a 30° angle is approximated using the equation. The nearest maximum error rounded to six decimal places is

(A) 2×10^{-6}

(B) 4×10^{-6}

(C) 8×10^{-6}

(D) 1×10^{-7}

6. Consider the following program segment.

```
10    INPUT A
20    B = 1
30    ITER = 1
40    ITER = ITER + 1
50    C = 0.5*(B + A/B)
60    B = C
70    GOTO 40
80    END
```

For A = 100 and ITER = 4, the value of C is most nearly

(A) 10.00

(B) 10.15

(C) 10.81

(D) 14.92

7. A matrix is shown.

$$\begin{vmatrix} 1 & 2 & -1 \\ 2 & 3 & 2 \\ 1 & -2 & -2 \end{vmatrix}$$

The determinant of the matrix is most nearly

(A) −7

(B) 10

(C) 17

(D) 24

8. Consider the following expression.

$$\lim_{x \to 27} \frac{\sqrt[3]{x} - 3}{x - 27}$$

The value of this expression is

(A) 1/3

(B) 1/9

(C) 1/27

(D) indeterminate

9. The three-number combination to open a safe is unknown. The lock keypad shows numbers from 0 to 9. The number of possible selections to open the combination lock is

(A) 9

(B) 81

(C) 729

(D) 1000

10. A factory employs 100 workers. Their wages vary according to the tasks they perform, as shown.

no. of workers	hourly wage
10	\$20.00
10	\$40.00
15	\$35.00
20	\$22.00
20	\$25.00
25	\$30.00

The median hourly wage of the workers is most nearly

(A) \$25.00/hr

(B) \$27.50/hr

(C) \$30.00/hr

(D) \$34.75/hr

11. A 500,000 ft^2 area is excavated to mine an ore. The site is expected to contain two types of materials: soft ore and rock. An engineering report estimates that the site contains 25% soft ore, and another report estimates that the site contains 55% soft ore. The unit costs are \$100/ft^2 to excavate soft ore and \$300/ft^2 to excavate rock. The expected excavation cost is most nearly

(A) \$100 million

(B) \$107 million

(C) \$110 million

(D) \$115 million

12. A data set consists of four points as shown.

x	y
2	9
3	11
5	15
9	22

Using least squares regression, the line equation that best fits the data is most nearly

(A) $y = 5.46 + 1.85x$

(B) $y = 5.0 + 2x$

(C) $y = 5.22 + 1.88x$

(D) $y = 4.95 + 1.89x$

13. The United States government uses a 1% annual exceedance probability (AEP) flood as the basis for the National Flood Maps. The probability that the peak discharge at a location in a 100-year flood plain will equal or exceed the 100-year flood level in the flood map in the next 100 years is most nearly

(A) 1%

(B) 33%

(C) 63%

(D) 100%

14. Which three of the following statements best define the Code of Ethics for Engineers? (Select the **three** that apply.)

(A) a set of guidelines that describes how a licensed engineer should behave professionally

(B) a set of aspirations that describes how a licensed engineer should behave professionally

(C) a set of rules that describes a licensed engineer's responsibilities to the public, clients, and other licensees

(D) a set of laws that describes how a licensed engineer must behave professionally

(E) a set of rules that incorporates criminal penalties

15. A licensed engineer's first and foremost responsibility, in performance of professional services, is to the

(A) clients

(B) employers

(C) customers

(D) public welfare

16. Which one of the following categories of original works by an author or inventor is NOT recognized as intellectual property?

(A) original works of authorship

(B) words, phrases, symbols, or designs made to identify a distinguishable source of a good or service

(C) unique inventions and discoveries

(D) secret techniques or processes that the inventor intends to use in business to obtain an economic advantage over competitors who do not know or use it

17. The four elements of sustainable construction or manufacturing are to

(A) reduce virgin resource consumption, use recyclable materials, protect nature, and focus on quality

(B) increase virgin resource consumption, eliminate imported materials, protect nature, and focus on quality

(C) increase virgin resource consumption, eliminate the use of recyclable materials, protect nature, and focus on quality

(D) reduce virgin resource consumption, eliminate the use of recyclable materials, protect nature, and focus on quality

18. Which of the following statements regarding greenhouse gas emissions is NOT correct?

(A) The carbon footprint method is a method used to measure and communicate the total amount of greenhouse gases emitted into the atmosphere both directly and indirectly in the production and delivery of goods and services.

(B) Raw materials, energy consumed, and transportation are the main elements used to calculate the carbon footprint of a product.

(C) A standard global method exists for calculating a carbon footprint.

(D) People or organizations that fully offset their carbon emissions by investing in green technology or sustainable projects are said to be carbon-neutral.

19. When providing design services for a construction project, which of the following is the standard of care that the courts traditionally assign to the engineer in responsible charge?

(A) The engineer should produce a set of perfect drawings.

(B) The engineer should meet the minimum requirement of the governing building code and no more.

(C) The engineer should use state-of-the-art methods that a reasonably prudent engineer would follow when designing similar projects in similar localities.

(D) The engineer is responsible for the job site safety of workers.

20. A community sells a property for \$24 in 1626, and the proceeds are invested at 7% interest compounded yearly. The value of the investment at the end of 2016 is most nearly

(A) \$70 million

(B) \$700 million

(C) \$7 billion

(D) \$7 trillion

21. Equipment selection has been narrowed to two options with the parameters shown. The expected rate of return for the equipment is 8%.

	equipment	
parameters	1	2
initial cost	$50,000	$75,000
annual maintenance	$15,000	$10,000
life expectancy (yr)	10	15
salvage value	$5000	$12,000

Which of the following statements is true?

(A) Equipment 1 is more economical because it saves nearly $8500.

(B) Equipment 1 is more economical because it saves nearly $1200.

(C) Equipment 2 is more economical because it saves nearly $8500.

(D) Equipment 2 is more economical because it saves nearly $1200.

22. A tollway authority at present charges $3.00 as a toll fee and is considering raising the toll fee. A total of 20,000 motorists use the road daily. It is estimated that for each $0.25 increase in the toll, 1000 fewer motorists will use the toll road. The toll fee increase that will maximize the authority's income is most nearly

(A) $1.00

(B) $1.50

(C) $1.75

(D) $2.25

23. Two alternatives for constructing a project are being considered. Alternative A costs $1 million initially and $100,000 in service and maintenance each year. Alternative B costs $1.8 million and $65,000 in service and maintenance each year. The project is expected to have no salvage value after a service life of 25 years, and the expected interest rate is zero. Which of the options should be used?

(A) alternative A

(B) alternative B

(C) either alternative A or B

(D) neither alternative A nor B

24. A piece of equipment is valued at $100,000. The equipment's value is depreciated at a rate of 10% per year. The book value after 10 years is most nearly

(A) $0

(B) $35,000

(C) $65,000

(D) $260,000

25. A contractor buys equipment at a cost of $150,000. After six years, its salvage value is estimated to be $10,000. Interest on the loan for the equipment is 8%. The annual cost is most nearly

(A) $23,000

(B) $29,000

(C) $31,000

(D) $34,000

26. A circuit is shown.

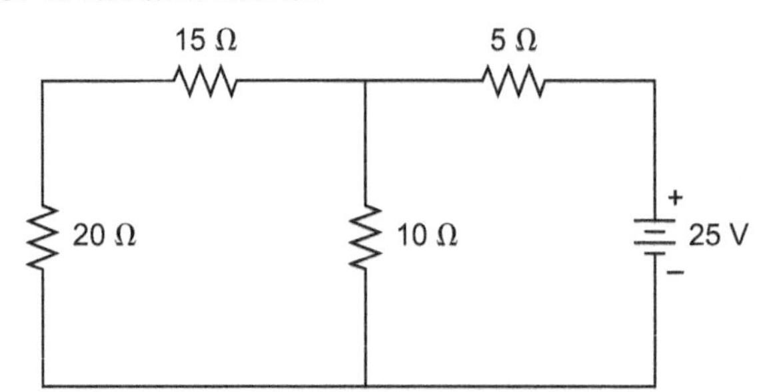

The current in the 10 Ω resistor is most nearly

(A) 0.43 A

(B) 1.53 A

(C) 1.96 A

(D) 2.50 A

27. A circuit is shown.

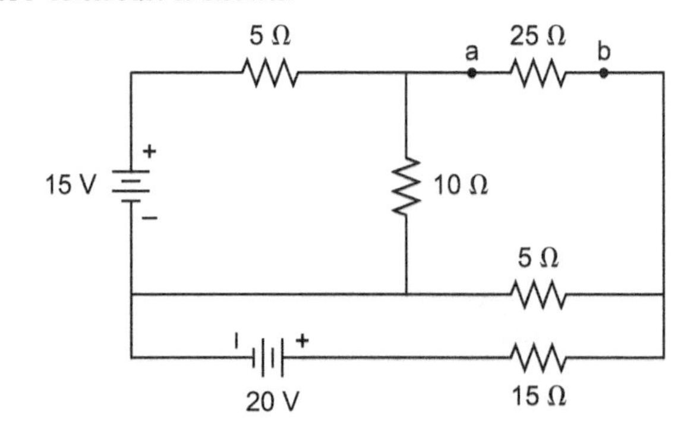

Model the circuit by a single source and a single resistor in series as shown.

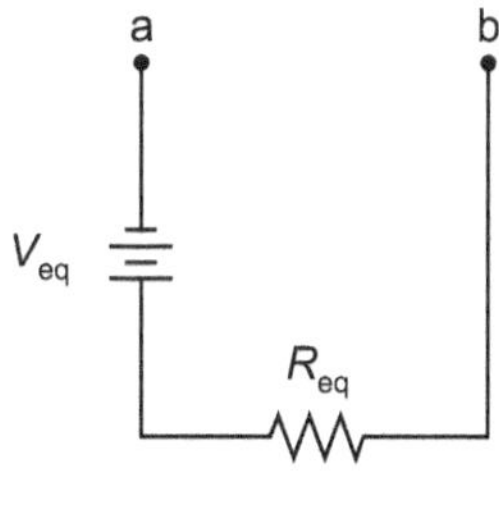

Equivalent circuit

The voltage across terminals a and b is most nearly

(A) 5 V

(B) 10 V

(C) 13 V

(D) 25 V

28. From the following statements, select all statements that are correct about Kirchhoff's current law (KCL).

(A) KCL is applicable to any closed path.

(B) KCL is applicable to any closed surface.

(C) KCL states that arithmetic sum of current at a node is zero.

(D) KCL states that algebraic sum of current at a node is zero.

(E) KCL requires that the clockwise direction of current should be assumed positive and counterclockwise direction should be assumed negative.

29. Two point charges are placed at points A and C. A third point charge is placed at point B in line with the other two so that it is not subject to any net force from the other two point charges. The electrical charge units are in coulombs, as shown.

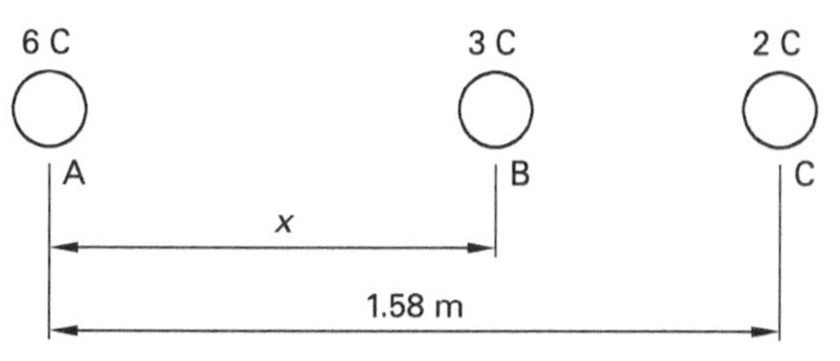

The distance AB is most nearly

(A) 44 cm

(B) 58 cm

(C) 100 cm

(D) 102 cm

30. An eight-pole alternator is running at 750 rpm and supplying power to a six-pole, three-phase induction motor. The motor has a full-load slip of 4%. The full-load speed of the motor is most nearly

(A) 720 rpm

(B) 750 rpm

(C) 780 rpm

(D) 960 rpm

31. A bolt is being pulled by two forces **A** and **B** as shown.

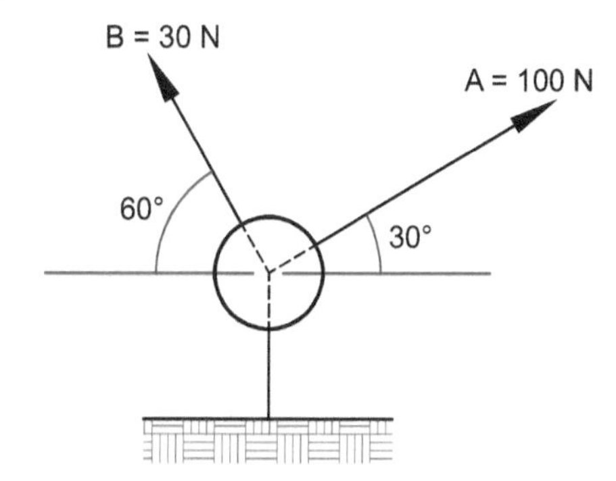

Which of the following diagrams best represents the resultant, **R,** graphically?

(A) 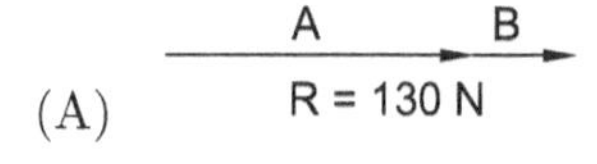

(B) 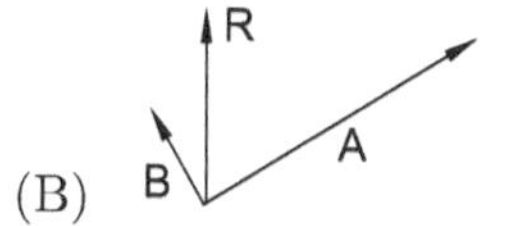

(C) 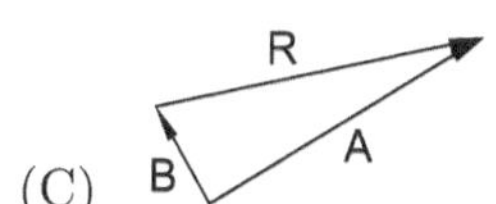

(D) 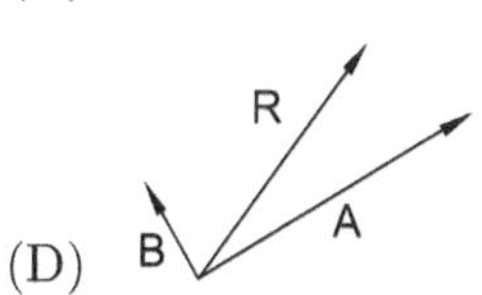

32. A moment couple is a system of

(A) two equal moments at the ends of a beam

(B) two equal and opposite moments at the ends of a beam

(C) two equal and opposite parallel forces

(D) two equal and opposite concurrent forces

33. A concrete block weighing 1200 lbf needs leveling on one edge. A crowbar is used to lift the edge, as shown.

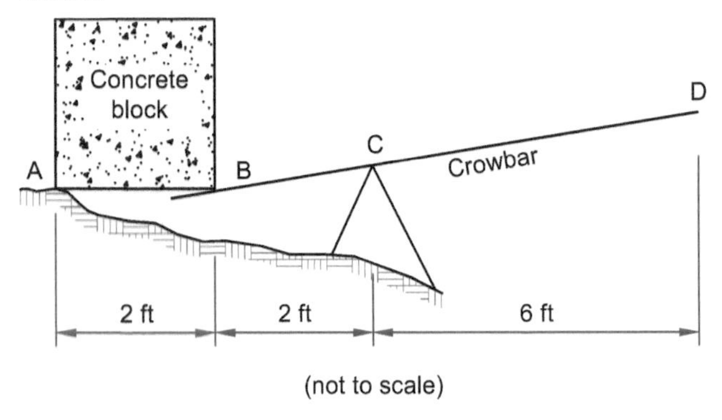

The downward force required at point D to lift the block is most nearly

(A) 100 lbf

(B) 200 lbf

(C) 600 lbf

(D) 800 lbf

34. A vertical pole is held by two stay cables, as shown.

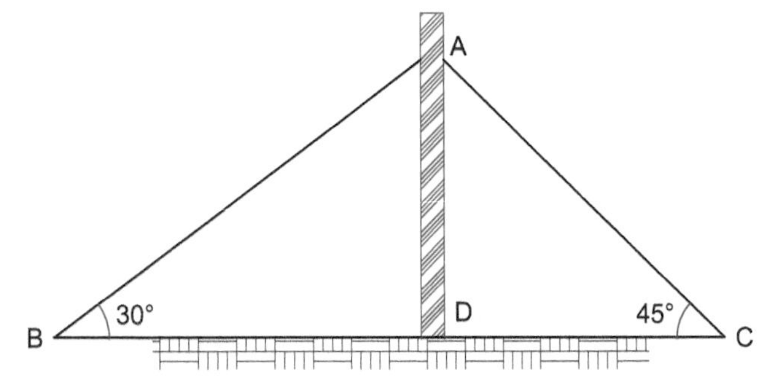

All connections are pinned. If the tension in cable AB is 100 N, the tension in cable AC is most nearly

(A) 100 N

(B) 120 N

(C) 150 N

(D) 180 N

35. The truss tower is shown. It has nine joints and 17 members. All joints are pinned, and the bars making up the X-bracings are not connected at their cross-points.

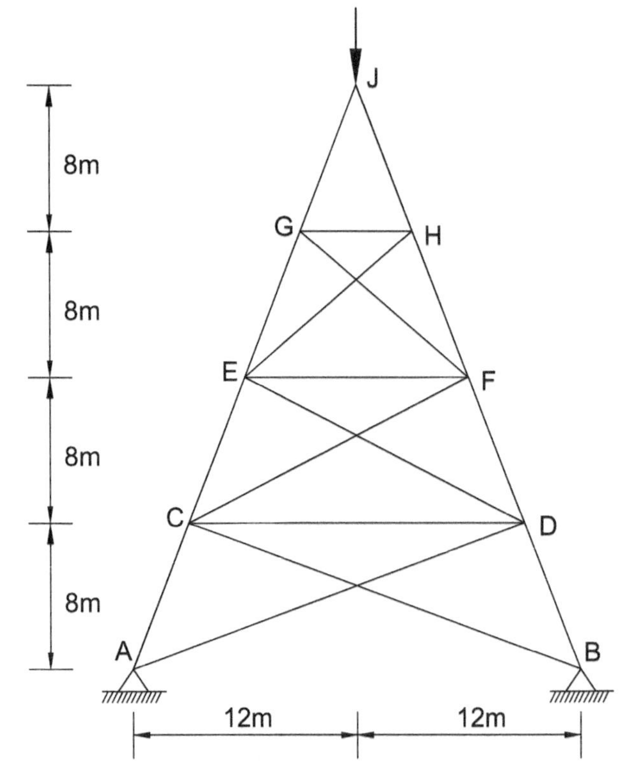

The total degree of indeterminacy of the tower is

(A) 1

(B) 2

(C) 3

(D) 4

36. Consider a 12 in radius circular disk. Its center is located at point A. A hole of 12 in diameter is cut from the disk. The center of the hole is located at point B. The distance between A and B is 6 in.

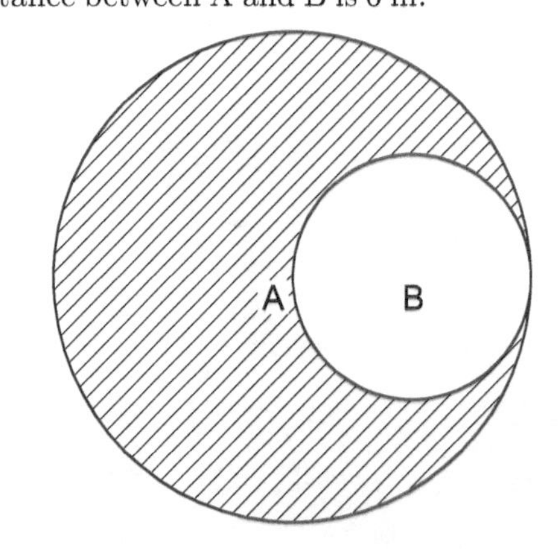

Where is the center of gravity of the composite disk located?

(A) at point A

(B) within segment AB

(C) left of point A, in line with segment AB

(D) somewhere else

37. A steel girder is made from three welded plates as shown.

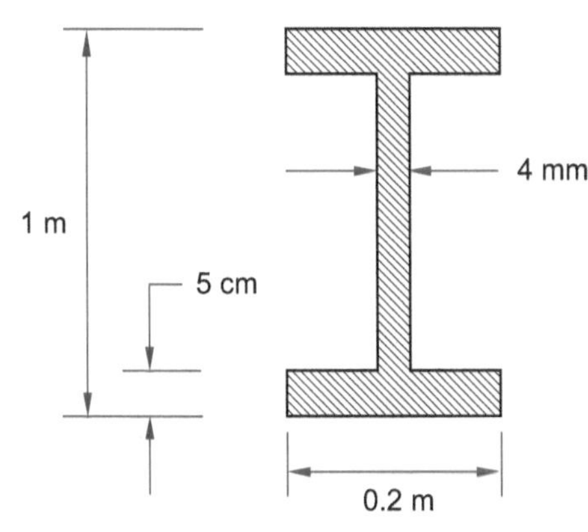

The moment of inertia about the major axis of the girder is most nearly

(A) $0.5 \times 10^{-3}\ \text{m}^4$

(B) $3.3 \times 10^{-3}\ \text{m}^4$

(C) $16.6 \times 10^{-3}\ \text{m}^4$

(D) $4.76 \times 10^{-3}\ \text{m}^4$

38. Consider the belt friction formula.

$$F_1 = F_2 e^{\mu\theta}$$

Which of the following statements are true for the equation shown?

(A) The force F_1 is always larger than F_2.

(B) The force F_2 is always larger than F_1.

(C) The formula can be applied to friction problems where the angle of contact exceeds 360°.

(D) The formula is inapplicable to friction problems involving band brakes.

(E) The formula is inapplicable to friction problems involving ropes wrapped around a capstan that remains fixed.

39. Consider the ladder shown.

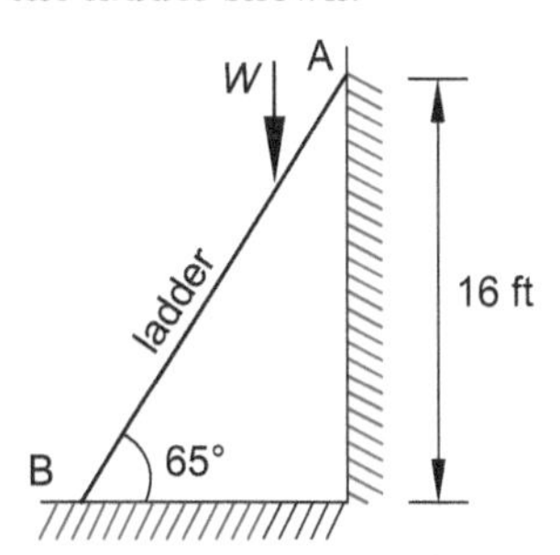

Assume zero friction between the ladder and wall. The minimum coefficient of friction needed at the floor for a 200 lbf person to safely use the ladder is most nearly

(A) 0.2

(B) 0.3

(C) 0.4

(D) 0.5

40. A cantilever is carrying a point load as shown.

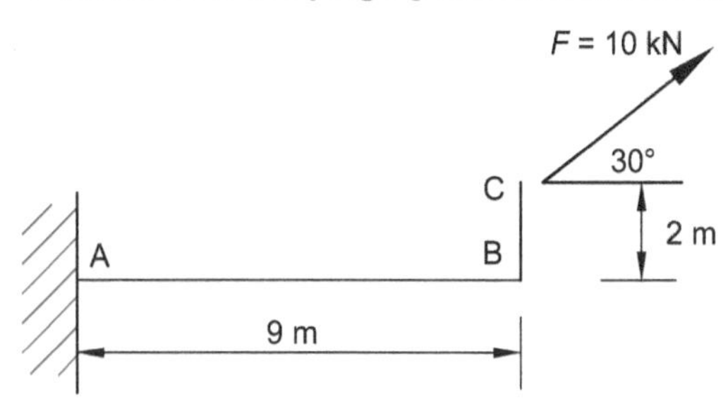

The moment at support A is most nearly

(A) 24 kN·m (ccw)

(B) 28 kN·m (ccw)

(C) 36 kN·m (ccw)

(D) 49 kN·m (cw)

41. A stone is dropped from a cliff top. The stone is seen hitting the ground 2.9 s later. Neglect air resistance to the stone. The cliff height above ground is most nearly

(A) 12 m

(B) 32 m

(C) 41 m

(D) 55 m

42. A concrete block weighing 145 lbf is resting on a concrete floor. The bearing surface of the block is 11 in square. The coefficient of friction is 0.7. The minimum force needed to slide the block is most nearly

(A) 82.5 lbf

(B) 102 lbf

(C) 145 lbf

(D) 173 lbf

43. A constant torque of 0.1 N·m is applied to an initially stationary flywheel with a moment of inertia of 3.14 kg·m^2. At the end of its tenth revolution, the speed of the flywheel is most nearly

(A) 1 rad/s

(B) 2 rad/s

(C) 3.14 rad/s

(D) 31.4 rad/s

44. A car weighs 6000 lbf and is traveling east at 30 mph when it collides head on with a 15,000 lbf truck traveling west at the same speed. After the collision, the vehicles stick together. Momentum is conserved during the collision. The vehicles' common velocity after the crash is most nearly

(A) 8 mph to the west

(B) 10 mph to the west

(C) 13 mph to the west

(D) 14 mph to the west

45. A 4 lbf weight stretches a spring 6 in over its natural length. The undamped natural frequency of the spring vibration is most nearly

(A) 6 rad/sec

(B) 2π rad/sec

(C) $4\sqrt{3}$ rad/sec

(D) 8 rad/sec

46. A baseball with a mass of 149 g travels at 30 m/s and is caught by a player and brought to rest in 0.1 sec. The average force applied to bring the ball to rest is

(A) 9.81 N

(B) 44.7 N

(C) 149 N

(D) 194 N

47. An asphalt pavement roller has a mass of 10,000 kg. Its wheels have a radius of gyration of 0.5 m and a mass of 5000 kg each. The diameter of each wheel is 1.1 m. The roller moves at a speed of 10 km/h. The total kinetic energy of the roller is most nearly

(A) 15,000 N·m

(B) 30,000 N·m

(C) 40,000 N·m

(D) 70,000 N·m

48. A vehicle weighing 10,000 lbf traveling at 60 mph comes to a stop without skidding. The wheelbase and center of gravity of the vehicle are shown.

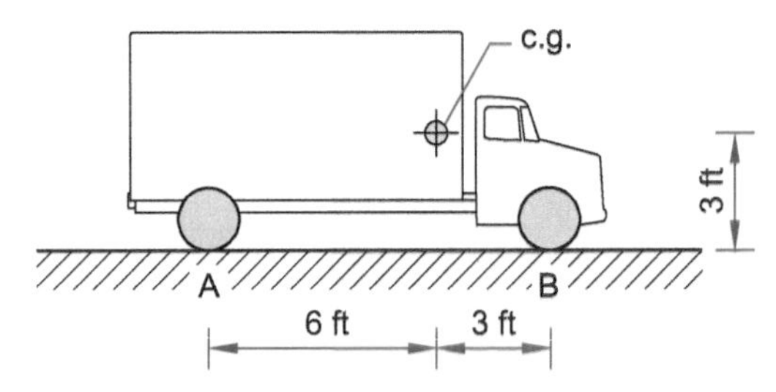

The deceleration is 13 ft/sec^2. The coefficient of friction, μ, between the road and the vehicle is most nearly

(A) 0.2

(B) 0.4

(C) 0.8

(D) 1.0

49. An impulse force acts on a 100 kg mass sitting on an inclined plane with a slope of 3/4 as shown.

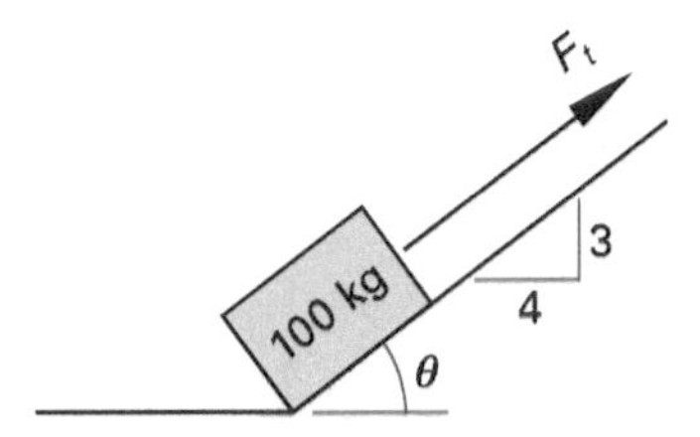

The impulse force is zero at 0 s and increases by 400 N/s.

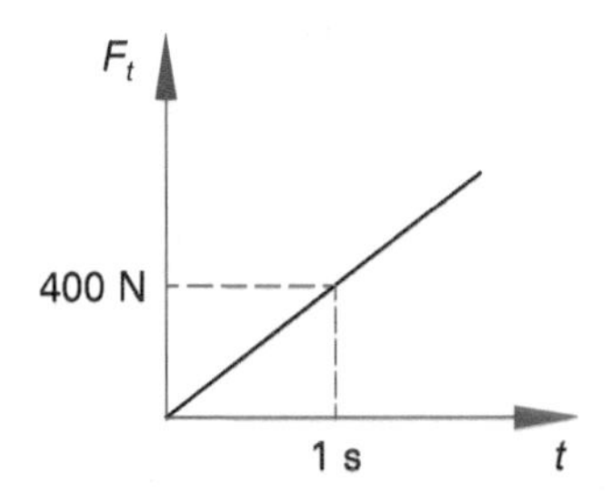

The coefficient of friction between the block and the sloping surface is 0.2. The time it takes for the block to start moving is most nearly

(A) 0 s

(B) 1.33 s

(C) 1.87 s

(D) 9.81 s

50. A three-rod mechanism A-B-C-D is shown.

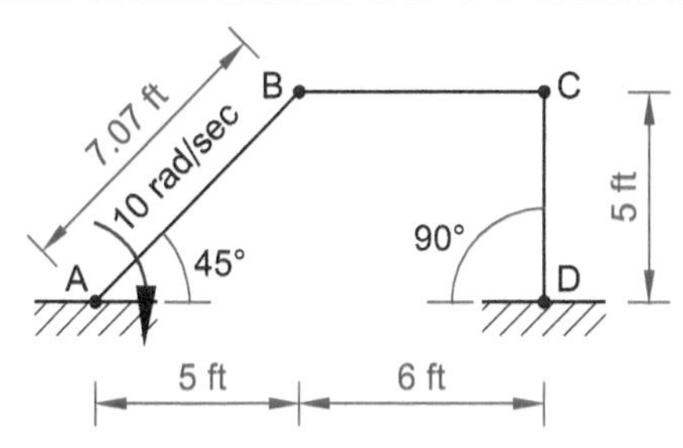

The rod AB is rotating at an angular speed of 10 rad/sec. The linear velocity at point B is most nearly

(A) 5.0 ft/sec

(B) 7.1 ft/sec

(C) 35 ft/sec

(D) 71 ft/sec

51. A steel bar is pulled by a 300 kN axial force. The bar is 1 m long and has a diameter of 3 cm. Its elongation is most nearly

(A) 1 mm

(B) 2 mm

(C) 5 mm

(D) 7 mm

52. A 1 in diameter bolt resists a 16 kips force as shown.

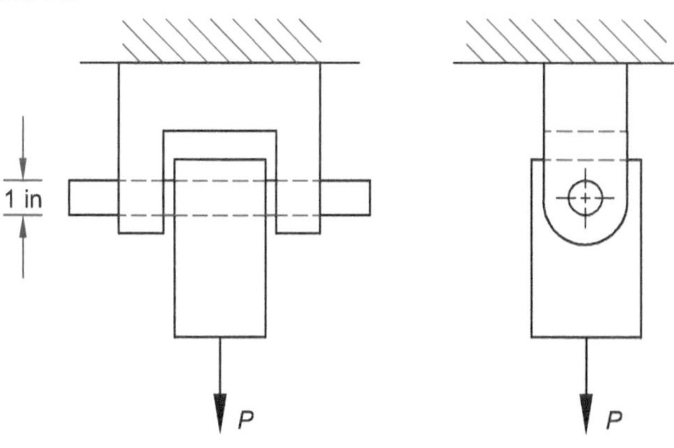

The shear stress in the bolt is most nearly

(A) 7.5 ksi

(B) 10 ksi

(C) 15 ksi

(D) 20 ksi

53. A 6 cm square, 2 m long steel bar is loaded concentrically. The bar's ends are pinned. Its buckling capacity is most nearly

(A) 53 kN

(B) 150 kN

(C) 530 kN

(D) 1500 kN

54. A metallic surveying tape is calibrated at 68°F. The tape is used in the field when the temperature is 98°F. The modulus of thermal expansion of the tape material is 0.0000065 in/in-°F. The measured distance is 3001.20 ft. The true distance is most nearly

(A) 2999.32 ft

(B) 3000.61 ft

(C) 3001.79 ft

(D) 3003.08 ft

55. A simply supported beam carries a uniformly distributed load (UDL) over its entire span. The UDL is replaced by a concentrated load, P, that equals the UDL in magnitude; the location of the load P coincides with the center of gravity (CG) of the UDL. Which forces remain the same for the beam under both loadings?

(A) External forces

(B) Internal forces

(C) Deflections

(D) None of the above

56. A simply supported beam is shown.

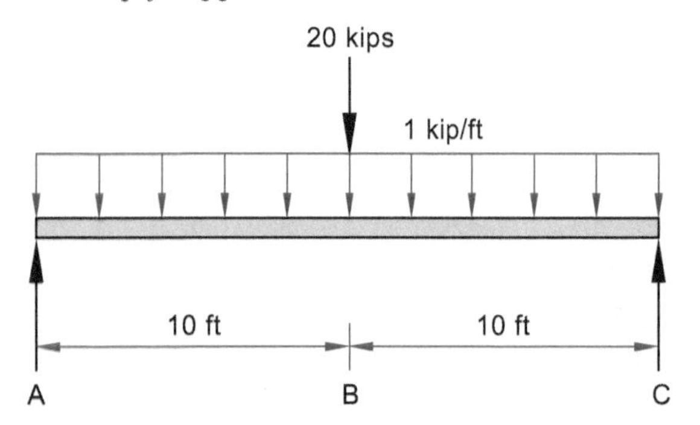

Which of the following is the corresponding shear force diagram?

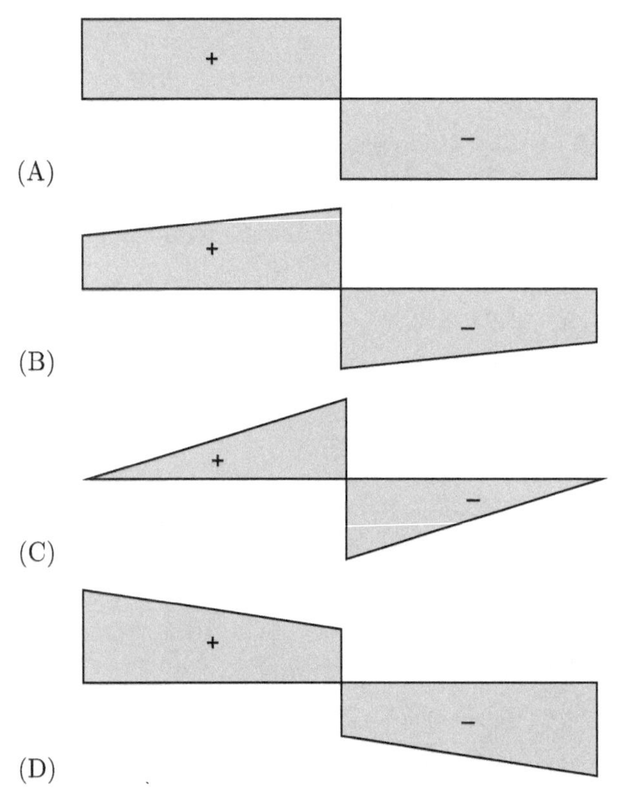

57. Consider a two-span beam as shown. The beam has two equal spans which carry a uniformly distributed load (UDL) of 3 N/m. The maximum moment in a propped cantilever under a UDL is $wL^2/8$.

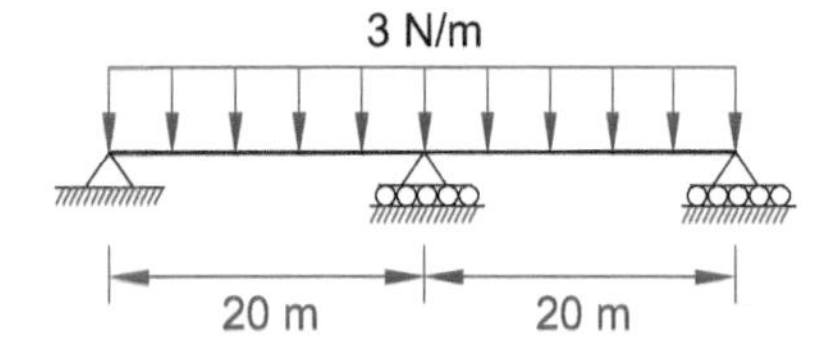

The bending moment of the beam at its central support is

(A) 0 N·m

(B) 120 N·m

(C) 150 N·m

(D) 600 N·m

58. A 1 in diameter and 3 ft long bar specimen of a low steel alloy with a yield stress of 36 ksi is tested under tension to an elongation of 0.20 in and then unloaded. The steel behavior during the test is elastic-plastic. The permanent elongation of the bar is most nearly

(A) 0 in

(B) 0.0452 in

(C) 0.155 in

(D) 0.275 in

59. A 200 mm diameter steel shaft is 6 m long and bears a uniformly distributed load of 250 N/m as shown.

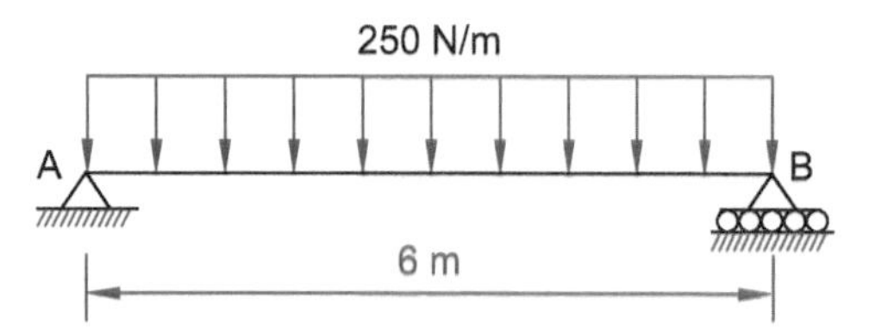

Neglecting self-weight, the shaft's maximum bending stress is most nearly

(A) 0.0014 MPa

(B) 0.14 MPa

(C) 1.4 MPa

(D) 14 MPa

60. The radius of a Mohr's circle is equal to

(A) sum of two principal stresses

(B) difference of two principal stresses

(C) one-half the sum of two principal stresses

(D) one-half the difference of two principal stresses

61. A point in a thin metallic plate is under biaxial stress of 100 ksi in the x-direction and 50 ksi in the y-direction, as shown.

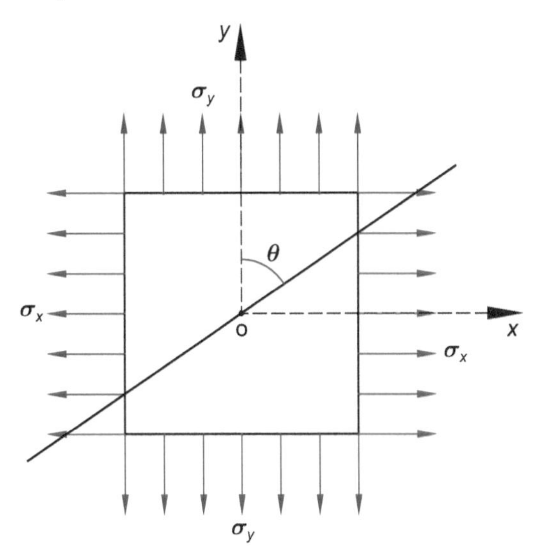

The shear stress at 30° from the minor axis is most nearly

(A) 12 ksi

(B) 22 ksi

(C) 25 ksi

(D) 76 ksi

62. Which statement regarding the standard oxidation potential for corrosion reaction of a metal is NOT correct?

(A) It signifies the rate at which corrosion is taking place.

(B) It is expressed in volts.

(C) A method is available to measure its absolute value.

(D) A metal with a higher oxidation potential corrodes at a higher rate.

63. The stress-strain diagram of a material is shown.

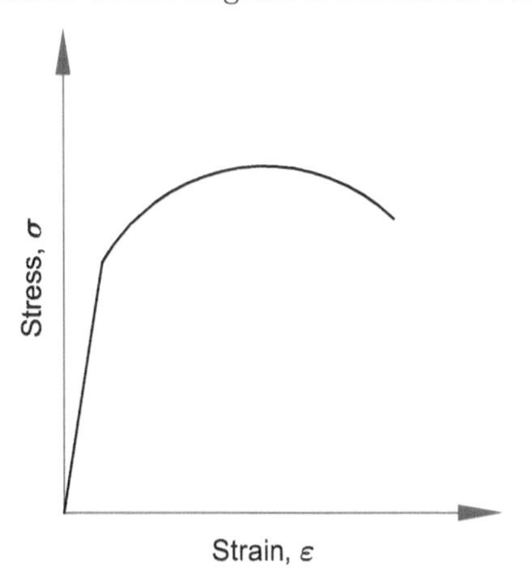

The yield strength of the material is 60 ksi, and the associated strain is 0.002. The modulus of resilience of the material is most nearly

(A) 12 psi

(B) 60 psi

(C) 120 psi

(D) 600 psi

64. A high-strength steel element with an ultimate strength of 1500 MPa is subjected to an alternating stress. No test data is available. The estimated endurance of the element is most nearly

(A) 70 MPa

(B) 140 MPa

(C) 700 MPa

(D) 1500 MPa

65. Which of the following statements regarding the uniaxial stress-strain relationship of metals is NOT correct?

(A) The slope of the linear portion of the curve equals the modulus of elasticity.

(B) For nonlinear materials, the stress at 0.2 percent strain offset is called the yield strength.

(C) The engineering stress is defined as the load divided by the initial cross-sectional area.

(D) The true stress is defined as the average of the yield stress and the ultimate stress.

66. A fiberglass composite is manufactured using the following components.

	e-glass fibers	epoxy
volume	70%	30%
modulus of elasticity	70.5 GPa	6.85 GPa

A sample of the material is taken and loaded uniaxially. Assuming all fibers in the sample are aligned parallel to the line of load, the percentage of load carried by the matrix is most nearly

(A) 4.0%

(B) 12%

(C) 21%

(D) 38%

67. A solution is composed of iron and 0.28% carbon. What, most nearly, is the amount and composition of primary α phase just above the eutectoid isotherm?

(A) 0%

(B) 35%

(C) 65%

(D) 77%

68. Which of the following combinations of metals is used to manufacture Y alloy?

(A) copper and zinc

(B) copper and tin

(C) copper, tin, and zinc

(D) aluminum, nickel, copper, and magnesium

69. Oil flows through a pipe at a velocity of 10 m/s. The specific gravity is 2.0, and the kinematic viscosity is 0.0003 m^2/s. For a 2.54 cm diameter pipe, the Reynolds number of the oil is most nearly

(A) 2.5

(B) 280

(C) 850

(D) 10,000

70. Water flows from tank A to tank B using the pipe system as shown.

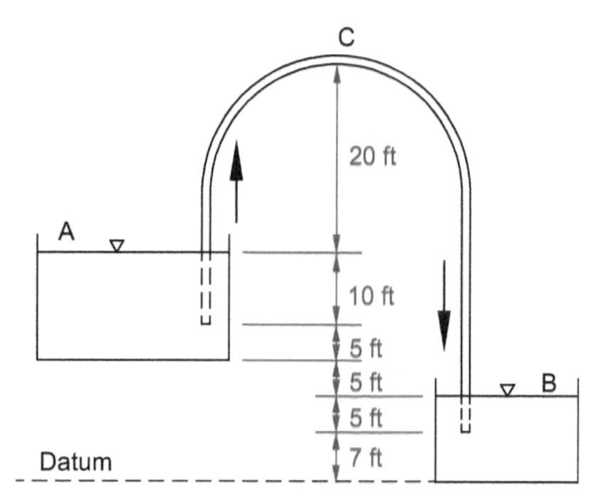

What is the difference in the hydraulic heads of the tanks?

(A) 15 ft

(B) 20 ft

(C) 40 ft

(D) 45 ft

71. Water enters the piping system at point A in a 10 in diameter pipe at the volumetric flow rate of 10 ft^3/sec, as shown. All pipes are laid horizontal.

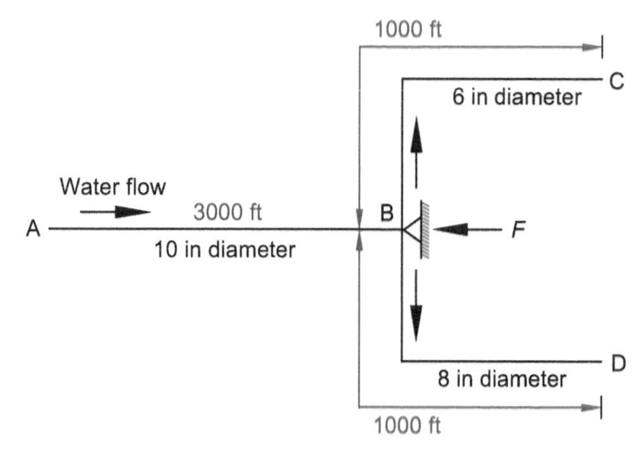

A rigid block is placed at joint B to stop pipe AB from moving. The force, F, exerted by water against the block is most nearly

(A) 110 lbf

(B) 350 lbf

(C) 2200 lbf

(D) 12,000 lbf

72. A 12 in diameter pipe is used to carry water at a flow rate of 3 ft/sec from reservoir A to reservoir B, as shown. Separation of dissolved gases occurs at an absolute pressure of 8 ft of water, and the change in the water level in each tank is negligible. Assume full pipe flow, an entry loss of 1.0 ft, and an exit loss of 0.1 ft. The pipe friction loss between points A and C is 15 ft, and the barometric pressure is 30 in of mercury (or 34 ft of water).

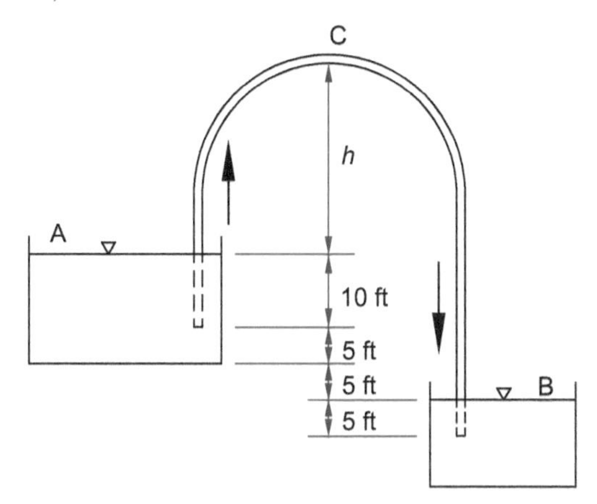

The maximum height, h, that can be used for siphoning is most nearly

(A) 0 ft

(B) 5.0 ft

(C) 9.8 ft

(D) 14 ft

73. A spherical air balloon having a 1 m diameter is propelled up into the air at a speed of 3.4 m/s. The air temperature is 40°C, and the air properties are

$$\text{density} = 1.12\ \text{kg/m}^3$$
$$\text{kinematic viscosity} = 1.70 \times 10^{-5}\ \text{m}^2/\text{s}$$

The drag force on the balloon is most nearly

(A) 1.0 N

(B) 2.0 N

(C) 3.0 N

(D) 4.0 N

74. Air is flowing from a reservoir to the atmosphere through a nozzle as shown.

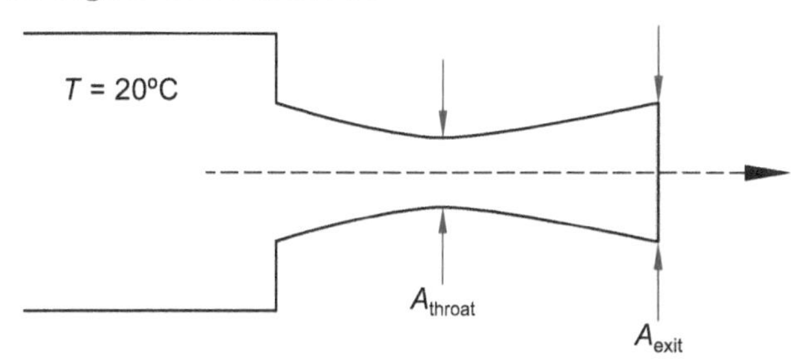

At the throat, the nozzle area is 1.20 in^2, and the Mach number is 1. Assume isentropic flow and ideal gas relations. The air velocity at the exit point is specified to be 5 Ma. What, most nearly, is the nozzle area at the exit?

(A) 3.5 in^2

(B) 12 in^2

(C) 25 in^2

(D) 30 in^2

75. A torque of 2 N·m is needed to rotate the cylinder shown at 1000 rad/s.

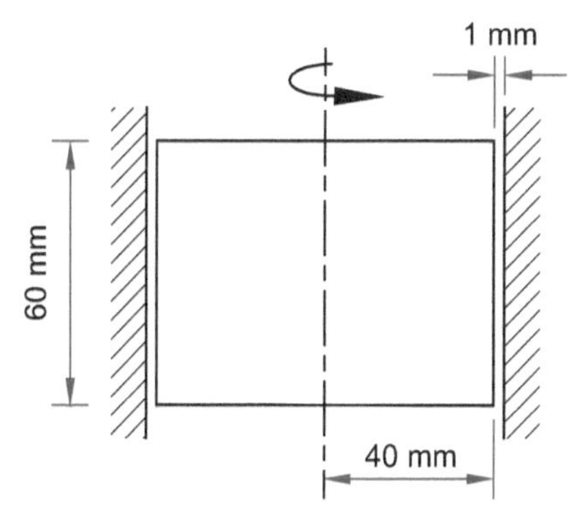

Assuming thin Newtonian film and linear velocity profile, the viscosity at the interface is most nearly

(A) 0.041 N·s/m^2

(B) 0.083 N·s/m^2

(C) 0.48 N·s/m^2

(D) 3300 N·s/m^2

76. A pumping system is needed to lift water 100 ft from one reservoir to the other as shown.

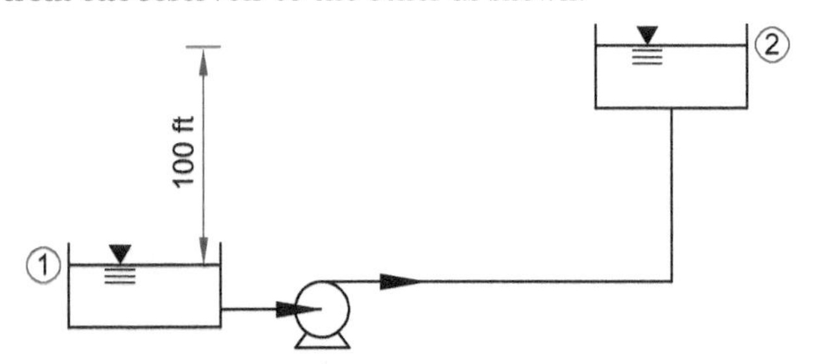

The reservoirs are connected by 1000 ft long, 6 in diameter pipe. The pipe's friction factor, f, is 0.0198. The system head curve is a parabola of the form

$$H = a + bQ^2$$

Using ft-sec units, what are the coefficients a and b, rounded to the nearest integer? ______ Enter your response in the blank.

77. A pump has the following operating characteristics.

Rotational speed (rpm)	1500
Efficiency	70%
Discharge (gpm)	670
Head (ft)	65

The same pump is required to deliver water at a head of 100 ft at the same efficiency. The required horsepower is most nearly

(A) 10 hp

(B) 20 hp

(C) 30 hp

(D) 60 hp

78. A nozzle of 1 in diameter delivers a stream of water at a velocity of 300 ft/sec impinging on a vane which moves at a velocity of 100 ft/sec and deflects the jet 60°.

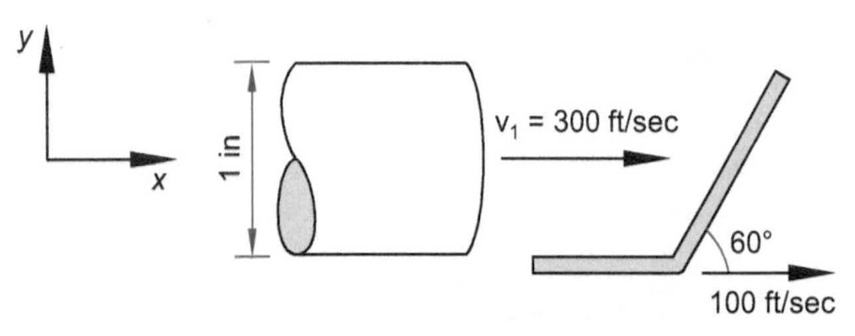

The horsepower of the turbine is most nearly

(A) 1.6 hp

(B) 3.2 hp

(C) 32 hp

(D) 60 hp

79. Which one of the following is an extensive property of a single component system in thermodynamics?

(A) volume

(B) pressure

(C) specific volume

(D) temperature

80. 2 kg of steam is contained in a rigid container under 0.2 MPa pressure at 300°C. In order to double the pressure, the energy that should be added is most nearly

(A) 1970 kJ

(B) 2810 kJ

(C) 3190 kJ

(D) 3890 kJ

81. A condenser inlets 10 kg/s of saturated water vapor at 1 MPa from a turbine and cools it by heating water from 20°C to 30°C. The pressure remains constant. What is the mass flux in kg/s, rounded to the nearest integer? ______ Enter your response in the blank.

82. 1 lbm of steam is expanded isentropically from an initial pressure of 1000 psia and 900°F to atmospheric pressure. What are its drop in enthalpy, in Btu, and the constant moisture, as a percentage? Round your answer to the nearest integer. ______ Enter your response in the blank.

83. Choose the definitions of carburetor and piston from the statements described below.

(A) It is a device to keep the engine speed, more or less, uniform at all load conditions.

(B) It is a device that mixes air and fuel in an appropriate air-fuel ratio for combustion.

(C) It is a device for firing the explosive mixture in an internal combustion engine.

(D) It is a device used to change phase in a cycle.

(E) It is a device to transfer force from expanding gas in the cylinder.

Carburetor definitions:
Piston definitions:

84. A heat engine absorbs 5000 J of energy from a hot reservoir at 500K and rejects 4000 J of heat energy to a cold reservoir at 300K. What is the efficiency (as a percentage) with respect to the Carnot cycle? ______ Enter your response in the blank.

85. Refrigerant R-410A is used in a cycle which operates between 30 psia and 200 psia. The refrigerant leaves the condenser as saturated liquid and enters the compressor as saturated vapor. What is the heat rejected in the condenser rounded to the nearest 10 Btu/lbm? ____ Enter your response in the blank.

86. 8 kg of oxygen and 7 kg of nitrogen are mixed. Both gases have a temperature of 57°C and a pressure of 1.1 bar. Both temperature and pressure are kept constant after mixing. What is the change in specific entropy in oxygen, in kJ/K? Round your answer to the nearest 0.01 kJ/K. ______ Enter your response in the blank.

87. Atmospheric air at 29.921 in of mercury pressure has 80°F dry-bulb temperature and 60°F wet-bulb temperature. What is the relative humidity as a percentage and the dew point in °F rounded to the nearest integer? ____ Enter your response in the blank.

88. 1 kg of methane is burned in air. What is the mass in kilograms of the products of combustion rounded to the nearest integer? ______ Enter your response in the blank.

89. Which of the following statements is correct?

(A) Law of cooling: Heat transferred from a hot body to a cold body is directly proportional to the surface area and the temperature gradient between the two bodies.

(B) Fourier's law states

$$\dot{Q} = kA\frac{dT}{dx}$$

The term x in the equation stands for the distance between two bodies.

(C) Convection is a process of heat transfer from a hot body to cold body, in a straight line, without affecting the intervening medium.

(D) Thermal conductivity of solid metals decreases with an increase in temperature.

90. The table shows design data for a plane wall.

Outside wall face temperature	–20°C
Inside wall face temperature	20°C
Wall thickness	15 cm
Wall area	2.5 m tall × 6 m long
Coefficient of thermal conductivity	1 W/m·K

What is the heat loss? (Round your answer to the nearest whole number of kilowatts.) ______ Enter your response in the blank.

91. From the five options, drag and drop the correct option in the blank space provided in the following statements.

1. ______ describes a process wherein heat flows from one particle of the body to another in the direction of fall of temperature, while the particles remain in fixed position relative to each other.

2. ______ describes a process wherein heat flows from one particle of the body to another in the direction of fall of temperature, while the particles move relative to each other.

3. ______ describes the change in heat flow rate as the difference in the fourth powers of the absolute temperatures of the object and of its environment.

(A) Conduction

(B) Convection

(C) Radiation

(D) Newton's law of cooling

(E) Stefan-Boltzmann law

92. The table shows design data for a cylindrical oven.

Inside diameter	2 cm
Heating element length	1.98 m
Inside temperature	1000°C
Wall temperature	500°C
Gray body reflectivity factor	0.15

What, most nearly, is the heat loss?

(A) 2.4 kW

(B) 14 kW

(C) 140 kW

(D) 1600 kW

93. A thermocouple is used to measure temperature in a gas pipe. The thermocouple is spherical in shape with a diameter of 9 mm. Its thermal conductivity is 20 W/m·K. The convection heat-transfer coefficient of the gas is 400 W/m^2·K. What, most nearly, is its Biot number?

(A) 0.01

(B) 0.03

(C) 0.50

(D) 1.00

94. Which of the following is NOT a thermodynamics property?

(A) Heat

(B) Pressure

(C) Temperature

(D) Specific volume

95. A metallic pipe carries steam at 200°C. The heat transfer data is

Pipe external diameter = 150 mm

Insulation thickness around pipe = 50 mm

Insulating material conductivity = 0.2 W/m·K

Temperature at outer surface of insulation = 85°C

What, most nearly, is the heat loss per meter per minute?

(A) 7100 J

(B) 17,000 J

(C) 37,000 J

(D) 97,000 J

96. Which one of the following statements is INCORRECT?

(A) The vacuum gauge pressure is the difference between the local atmospheric pressure and the absolute pressure.

(B) The pressure sensors are typically based on measuring the strain on a thin membrane due to an applied pressure.

(C) Transducer sensitivity is defined as the ratio of change in electrical signal magnitude to the change in magnitude of the physical parameter being measured.

(D) A Wheatstone bridge is said to be balanced only if all four of its resistances are equal.

97. A system represented by block diagrams is shown.

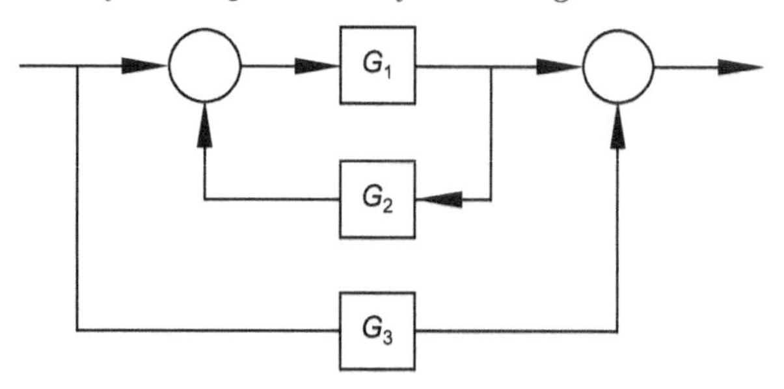

The system's total output is represented by the equation

$$G(s) = \frac{X}{1+Y} + Z$$

What do the terms X and Y stand for?

(A) X is G_1, and Y is G_2.

(B) X is G_1, and Y is G_1G_2.

(C) X is G_2, and Y is G_2G_3.

(D) X is G_3, and Y is G_1/G_2.

98. A motor weighing 400 lbf is supported by four springs, one in each corner. The stiffness of each spring is 1000 lbf/in. The motor is constrained to move vertically, and there is no imbalance in mass. What, most nearly, is the motor speed at which resonance will occur?

(A) 110 rpm

(B) 590 rpm

(C) 940 rpm

(D) 1800 rpm

99. A particle moves in a straight line for a distance measured to be $d = 100\text{ m} \pm 1\text{ m}$ during a time $t = 25.0\text{ s} \pm 3\text{ s}$. In meters per second, what is the uncertainty in measuring the particle's velocity rounded to one decimal place? ______ Enter your response in the blank.

100. A basic negative feedback system represented by block diagrams is shown.

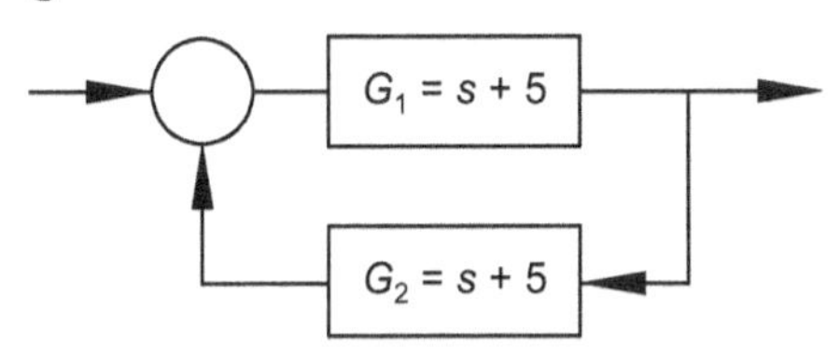

What is the system's time-domain transfer function?

(A) $e^{-t}\cos t$

(B) $e^{-t}\cos(5t)$

(C) $e^{-5t}\cos t$

(D) $e^{-5t}\cos(5t)$

101. A straight spur gear is designed to transmit 150 kW. Its properties are as follows.

Pitch circle diameter = 300 mm

Speed = 600 rpm

Pressure angle = 20°

The total load exerted on the gears is most nearly

(A) 14 kN

(B) 17 kN

(C) 63 kN

(D) 150 kN

102. A metallic crankshaft has the following properties.

Yield strength = 600 MPa

Ultimate strength = 1200 MPa

It is subjected to an alternating stress of 60 MPa with a mean stress of 60 MPa. Using the modified Goodman theory, the fatigue strength of the crankshaft is most nearly

(A) 63 MPa

(B) 140 MPa

(C) 480 MPa

(D) 540 MPa

103. A spring system is subjected to a point load as shown.

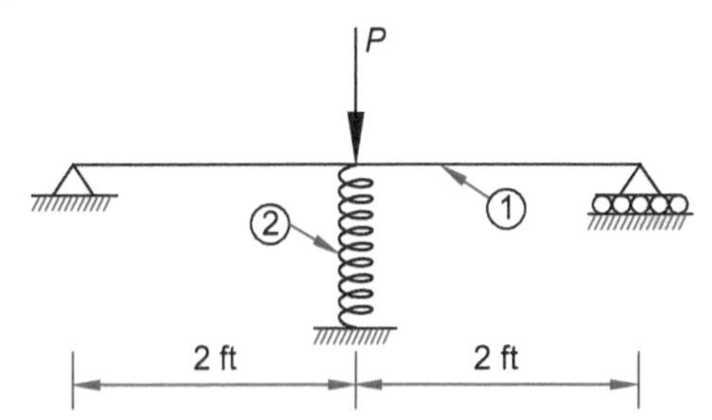

For element 1, the modulus of elasticity is 29,000 ksi, and the moment of inertia is 0.21 in^4. For element 2, the spring constant is 4 kips/in. The percentage of the applied load, P, being resisted by element 2 is most nearly

(A) 50%

(B) 60%

(C) 70%

(D) 80%

104. A closely coiled helical linear spring has a mean diameter of 12 in. It is made of 1 in diameter wire and is subjected to an axial load of 1 kip. The shear stress in the spring wire is most nearly

(A) 1 ksi

(B) 8 ksi

(C) 16 ksi

(D) 32 ksi

105. A steel thin-walled cylindrical pressure vessel contains a fluid at a gauge pressure of 125 psi. The maximum permissible steel stress is 24 ksi. The specified mean diameter of the cylinder is 4 ft. The minimum cylinder wall thickness is most nearly

(A) 1/8 in

(B) 1/4 in

(C) 1/2 in

(D) 1 in

106. A single deep-groove ball bearing is subjected to a radial force of 3.9 kN. The shaft rotates at a speed of 600 rpm. The specified design life is 30,000 hr. The minimum acceptable diameter of the shaft is 40 mm. What, most nearly, is the minimum basic load rating required for the bearings?

(A) 3,900

(B) 30,000

(C) 40,000

(D) 60,000

107. A square thread screw-jack has a single thread of 8 mm pitch on a mean thread radius of 30 mm. The screw-jack is used to raise a load of 500 kN. The coefficient of friction on the screw threads is 0.12. What, most nearly, is the magnitude of the torque?

(A) 1.2 kN·m

(B) 2.4 kN·m

(C) 6.0 kN·m

(D) 40 kN·m

108. The coefficient of friction between a belt and its pulley is 0.24, and their angle of contact is 180°. The tension in the drive side of the belt is 200 lbf. What, most nearly, is the tension in the belt side resisting the impending motion?

(A) 48 lbf

(B) 94 lbf

(C) 110 lbf

(D) 200 lbf

109. Two 1/2 in thick plates are joined by oversized rivets as shown. The rivets are designed to transfer a tensile load that either plate can carry.

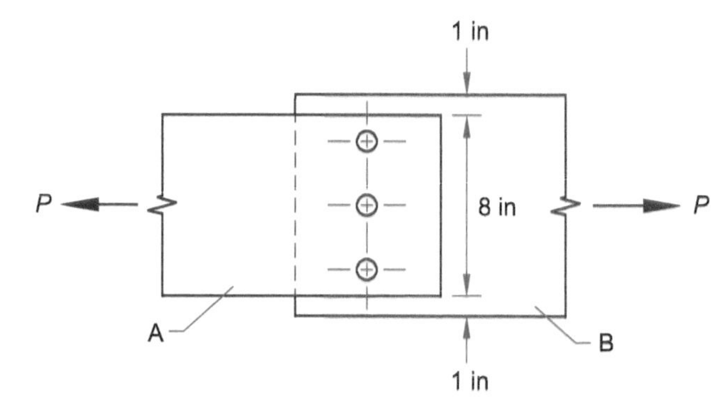

The assembly's most likely failure mode is by

(A) pure shear

(B) rupture

(C) crushing of rivets

(D) block shear

110. A solid shaft has a nominal diameter of 20 mm. Its lower and upper deviations are 0.04 mm and 0.03 mm. For clearance fit, what is the maximum basic diameter of the shaft?

(A) 20.01 mm

(B) 20.03 mm

(C) 20.04 mm

(D) 20.07 mm

STOP!

DO NOT CONTINUE!

This concludes the examination. If you finish early, check your work and make sure that you have followed all instructions. After checking your answers, you may turn in your examination booklet and answer sheet and leave the examination room. Once you leave, you will not be permitted to return to work or change your answers.

Exam 2 Instructions

In accordance with the rules established by your state, you may use any approved battery- or solar-powered, silent calculator to work this examination. However, no blank papers, writing tablets, unbound scratch paper, or loose notes are permitted. Sufficient paper will be provided. The *NCEES FE Reference Handbook* is the only reference you are allowed to use during this exam.

You are not permitted to share or exchange materials with other examinees.

You will have six hours in which to work this session of the examination. Your score will be determined by the number of questions that you answer correctly. There is a total of 110 questions. All 110 questions must be worked correctly in order to receive full credit on the exam. There are no optional questions. Each question is worth 1 point. The maximum possible score for this section of the examination is 110 points.

Partial credit is not available. No credit will be given for methodology, assumptions, or work written on scratch paper.

Record all your answers on the Answer Sheet. If you change your answer, make sure your final answer is clearly indicated. Select only one answer per question unless instructed otherwise. Some questions will require you to select more than one correct answer, select a point on a graphic, move items to their correct locations in a table or figure, or enter your response in a blank space. All questions have the same point value regardless of type.

If you finish early, check your work and make sure that you have followed all instructions. After checking your answers, you may submit your answers and leave the examination room. Once you leave, you will not be permitted to return to work or change your answers.

When permission has been given by your proctor, you may begin your examination.

Name: ______________________________

Last First Middle Initial

Examinee number: ______________

Examination Booklet number: ______________

Fundamentals of Engineering Examination

Exam 2

Exam 2 Answer Sheet

111. (A) (B) (C) (D)
112. (A) (B) (C) (D)
113. (A) (B) (C) (D)
114. (A) (B) (C) (D)
115. (A) (B) (C) (D)
116. (A) (B) (C) (D)
117. (A) (B) (C) (D)
118. (A) (B) (C) (D)
119. (A) (B) (C) (D)
120. (A) (B) (C) (D)
121. (A) (B) (C) (D)
122. (A) (B) (C) (D)
123. (A) (B) (C) (D)
124. (A) (B) (C) (D) (E)
125. (A) (B) (C) (D) (E)
126. (A) (B) (C) (D)
127. (A) (B) (C) (D)
128. (A) (B) (C) (D)
129. (A) (B) (C) (D)
130. (A) (B) (C) (D)
131. (A) (B) (C) (D)
132. (A) (B) (C) (D)
133. (A) (B) (C) (D)
134. (A) (B) (C) (D)
135. (A) (B) (C) (D)
136. (A) (B) (C) (D)
137. (A) (B) (C) (D)
138. ____________
139. (A) (B) (C) (D)
140. (A) (B) (C) (D)
141. (A) (B) (C) (D)
142. (A) (B) (C) (D)
143. (A) (B) (C) (D)
144. (A) (B) (C) (D)
145. (A) (B) (C) (D)
146. (A) (B) (C) (D)
147. (A) (B) (C) (D)
148. (A) (B) (C) (D) (E)
149. (A) (B) (C) (D)
150. (A) (B) (C) (D)
151. (A) (B) (C) (D)
152. (A) (B) (C) (D)
153. (A) (B) (C) (D)
154. (A) (B) (C) (D)
155. (A) (B) (C) (D)
156. (A) (B) (C) (D)
157. (A) (B) (C) (D)
158. (A) (B) (C) (D)
159. (A) (B) (C) (D)
160. (A) (B) (C) (D)
161. (A) (B) (C) (D)
162. (A) (B) (C) (D)
163. (A) (B) (C) (D)
164. (A) (B) (C) (D)
165. (A) (B) (C) (D)
166. (A) (B) (C) (D)
167. (A) (B) (C) (D)
168. (A) (B) (C) (D)
169. (A) (B) (C) (D)
170. (A) (B) (C) (D)
171. (A) (B) (C) (D)
172. (A) (B) (C) (D)
173. (A) (B) (C) (D)
174. (A) (B) (C) (D)
175. (A) (B) (C) (D)
176. (A) (B) (C) (D)
177. (A) (B) (C) (D)
178. ____________
179. (A) (B) (C) (D)
180. ____________
181. (A) (B) (C) (D)
182. (A) (B) (C) (D)
183. (A) (B) (C) (D)
184. (A) (B) (C) (D)
185. (A) (B) (C) (D)
186. (A) (B) (C) (D)
187. (A) (B) (C) (D)
188. (A) (B) (C) (D)
189. (A) (B) (C) (D)
190. (A) (B) (C) (D)
191. (A) (B) (C) (D)
192. (A) (B) (C) (D)
193. (A) (B) (C) (D)
194. (A) (B) (C) (D)
195. (A) (B) (C) (D)
196. (A) (B) (C) (D)
197. ____________
198. ____________
199. (A) (B) (C) (D)
200. (A) (B) (C) (D)
201. (A) (B) (C) (D)
202. (A) (B) (C) (D)
203. (A) (B) (C) (D)
204. (A) (B) (C) (D)
205. ____________
206. ____________
207. (A) (B) (C) (D)
208. ____________
209. (A) (B) (C) (D)
210. ____________
211. (A) (B) (C) (D)
212. (A) (B) (C) (D)
213. (A) (B) (C) (D)
214. (A) (B) (C) (D)
215. ____________
216. (A) (B) (C) (D)
217. (A) (B) (C) (D)
218. ____________
219. (A) (B) (C) (D)
220. ____________

Exam 2

111. For a geothermal project, the temperature increases from 20°C at the surface of the earth to 90°C at a depth of 2 km below the ground. Assuming a linear temperature gradient, the underground temperature at 3800 m below the ground surface is most nearly

(A) 113°C

(B) 133°C

(C) 153°C

(D) 173°C

112. A particle moves along a line so that its position at time t is

$$s(t) = \frac{t^2+1}{t+1}$$

The velocity at $t = 2.5$ s is most nearly

(A) 0.78

(B) 0.84

(C) 0.92

(D) 2.1

113. A differential equation is given.

$$y'' + 3y' + 2 = 0$$

Its general solution is

(A) $y = C_1e^{-2x} + C_2e^{-x}$

(B) $y = C_1e^{2x} + C_2e^{x}$

(C) $y = C_1e^{i2x} + C_2e^{ix}$

(D) $y = e^{x}(3\cos x + 2\sin x)$

114. Two matrices are given.

$$A = \begin{vmatrix} 1 & 2 \\ 5 & 4 \\ 3 & 7 \end{vmatrix}$$

$$B = \begin{vmatrix} 2 & 8 & 9 \\ 10 & 1 & 6 \end{vmatrix}$$

Matrices A and B are multiplied to obtain a product matrix C. Which statement about the matrix multiplication is true?

(A) The product matrix C will have two rows and three columns.

(B) The product matrix C will have three rows and two columns.

(C) The product matrix C will have three rows and three columns.

(D) The multiplication is not possible since the matrices A and B do not have the same number of rows and columns.

115. Consider the equation $x^7 - 100 = 0$. The zeroth root equals 2. Using Newton's method of root extraction, what is most nearly the value of x after the second iteration?

(A) 0.1429

(B) 1.9307

(C) 1.9308

(D) 1.9375

116. Consider the following program segment.

```
10    INPUT A
20    B= 1
30    ITER=1
40    ITER=ITER + 1
50    C=0.5*(B+A/B)
60    B=C
70    GOTO 40
80    END
```

What is the programing error in the algorithm?

(A) division by zero

(B) endless loop

(C) missing print statement

(D) both endless loop and missing print statement

117. What is the differential $\partial z/\partial s$ for $z = xy$ where $x = s^2 + t^2$ and $y = s/t$?

(A) $\dfrac{3s^2 + t^2}{t}$

(B) $\dfrac{s^2 + t^2}{t}$

(C) $\dfrac{s^2 + 3t^2}{t}$

(D) $\dfrac{s(s^2 + t^2)}{t}$

118. An integral is shown.

$$\int_{-2}^{1}(3x^2 + 2x - 9)\,dx$$

What is the value of the integral?

(A) −21

(B) −7

(C) 7

(D) 21

119. A crew of 5 workers is selected from a group of 8 male and 12 female workers. If the crew is made up of 1 man and 4 women, the number of ways the crew can be formed is

(A) 3960

(B) 11,900

(C) 15,500

(D) 95,000

120. Four concrete specimens are tested. Their strengths are 4450 psi, 4675 psi, 4898 psi, and 4120 psi. The sample standard deviation is most nearly

(A) 290 psi

(B) 330 psi

(C) 83,000 psi

(D) 110,000 psi

121. A 500,000 ft^2 area is excavated to construct a manufacturing plant. The foundation is expected to be partially on sandy silt and partially on rock. One geotechnical report estimates that the foundation is 25% sandy silt, and another report estimates that the foundation is 55% sandy silt. The unit cost to excavate sandy-silt type material is $\$100/ft^2$, and it costs $\$300/ft^2$ to excavate the rock to the specified depth. The owner gives the first report twice the weight of the second report. The foundation cost expected by the owner is most nearly

(A) \$100M

(B) \$107M

(C) \$110M

(D) \$115M

122. A data set consists of four points, as shown.

	x	y	xy	x^2
	2	9	18	4
	3	11	33	9
	5	15	75	25
	9	22	198	81
Σ	**19**	**57**	**324**	**119**

The mean standard error of estimate (MSE) is most nearly

(A) 0.061

(B) 0.092

(C) 2.3

(D) 4.4

123. A city's rainfall is modeled as a continuous random variable, x, and its probability, using the unit normal distribution, is described using the probability density function.

$$f(x) = \frac{1}{\sqrt{2\pi}}e^{-\frac{(x-15)^2}{2}}$$

The fraction of the year during which the rainfall is between 14 in and 16 in is most nearly

(A) 0.33

(B) 0.50

(C) 0.68

(D) 0.97

124. From the following statements, which describe the Code of Engineering Ethics? Select **all** that apply.

(A) a set of guidelines that describe how a licensed engineer should behave professionally

(B) a set of aspirations that describe how a licensed engineer should behave professionally

(C) a set of rules that describe a licensed engineer's responsibilities to the public, clients, and other licensees

(D) a set of laws that describe how a licensed engineer should behave professionally

(E) a set of rules that incorporate criminal penalties

125. An engineer develops a new idea to significantly reduce energy consumption in manufacturing plants. The engineer considers filing a patent application to the United States Patent and Trademark office (USPTO). Which categories of patents will the USPTO consider? Select **all** that apply.

(A) utility patent

(B) design patent

(C) plant patent

(D) energy patent

(E) efficiency patent

126. Which of the following statements are INCORRECT?

I. A mechanical engineer thinks of a new heat cycle to increase efficiency. Her idea is protected under the copyright laws.

II. A mechanical engineer thinks of a new heat cycle to increase efficiency and produces two-dimensional drawings. She keeps the drawings in a secure place. Her work is protected under the copyright laws.

III. A mechanical engineer examines the workings of a machine in a factory and builds a more efficient machine. She shows the new machine and her design drawings to the public. She is not guilty of copyright infringement.

IV. A mechanical engineer examines the workings of a machine and builds a more efficient machine. She shows the machine and her design to the public. Her design may be patented.

(A) I only

(B) I and III

(C) II and IV

(D) I, III, and IV

127. An engineering firm desires to reduce its risk arising out of its design work of its employees. Select the type of insurance that the firm should procure.

(A) general liability insurance

(B) comprehensive insurance

(C) engineer's hazard insurance

(D) negligence insurance

128. A licensee should approve and seal only those design documents that

(A) meet the minimum requirements of a building code

(B) safeguard the life, health, and welfare of the public

(C) are personally prepared by them

(D) require all green or environmentally friendly materials

129. A licensee makes a professional judgment that is overruled. The licensee believes that the life, health, property, or welfare of the public is endangered. Who should the licensee inform?

(A) the licensee's employer

(B) the licensee's client

(C) other authorities as may be appropriate

(D) A or B, and C

130. An engineer owns a design firm that generates $1 million per year in gross fees. He calculates that he receives $100,000 per year in profits after paying all employee salaries, benefits, and other expenses. Assume zero inflation and no growth. If the expected rate of return is 10%, the fair market price of the firm is most nearly

(A) $600,000

(B) $1,000,000

(C) $1,600,000

(D) $6,000,000

131. An engineering firm at present gets its drawings printed at a local blueprint facility. The firm is considering buying a new printer for printing its drawings in house. The printer costs \$40,000, and the annual operational and insurance cost is \$3000. Its life expectancy is 5 years with no salvage value. The printer is expected to generate \$15,000 per year in savings for the firm. The rate of return on the printer, as an annual percentage, is most nearly

(A) 9.0%

(B) 12%

(C) 15%

(D) 16%

132. A concrete mix requires two admixtures: a minimum of 3 doses but no more than 9 doses of admixture A, and a minimum of 8 doses of admixture B. The maximum sum of both doses is 16. The costs of admixtures A and B are \$4 and \$12 per dose, respectively. The least cost of admixtures to produce the mix is most nearly

(A) \$101

(B) \$108

(C) \$192

(D) \$216

133. A machine is covered by the manufacturer's warranty for the first year. After this, the repair cost is expected to be as follows.

cost	amount
at the end of year 2	\$1000
at the end of year 3	\$2000
at the end of year 4	\$3000
at the end of year 5	\$4000

The interest rate is 8%. The annual equivalent repair cost is most nearly

(A) \$1800

(B) \$2000

(C) \$2200

(D) \$2400

134. A design firm anticipates that it will need to upgrade its computers and drafting system in 5 years. The upgrade will cost \$500,000. The firm makes deposits every month into an account that earns 0.5% interest compounded monthly. The minimum amount of money the firm should deposit every month to save for the anticipated cost is most nearly

(A) \$7150

(B) \$8330

(C) \$8850

(D) \$100,000

135. A repair project will take 1 year to complete. It will cost \$100,000 now and \$50,000 at the beginning of the second year. The restoration will extend the life of the facility by 3 years after completion. At a 10% annual rate of return, the yearly minimum income the project should produce to make the project feasible is most nearly

(A) \$64,000

(B) \$65,000

(C) \$66,000

(D) \$67,000

136. A steel ring with 300 mm mean diameter is made of a 20 mm by 20 mm cross section. It is uniformly wound with 400 turns of wire that is 2 mm square in cross section. Steel's permeability is 8.8×10^{-3} H/m. The self-inductance of the coil is most nearly

(A) 0.2 H

(B) 0.6 H

(C) 8 H

(D) 9 H

137. A capacitor consists of two square aluminum plates 11 mm apart, each 100 mm square with sides parallel to each other. The relative permittivity of natural rubber is 7. The capacitance is most nearly

(A) 30 $\mu\mu$F

(B) 56 $\mu\mu$F

(C) 62 $\mu\mu$F

(D) 70 $\mu\mu$F

138. Two theorems are defined among the following statements.

(A) Whatever amount of current goes into a node of the circuit must come out of the node.

(B) The algebraic sum of the voltages is zero around any loop in the circuit.

(C) Any circuit can be reduced to a single current source with a parallel resistance.

(D) Any circuit can be reduced to a single voltage source in series with a resistance.

Move each applicable statement to the appropriate row.

1. Thevenin theorem
2. Norton theorem

139. A sawtooth waveform is shown.

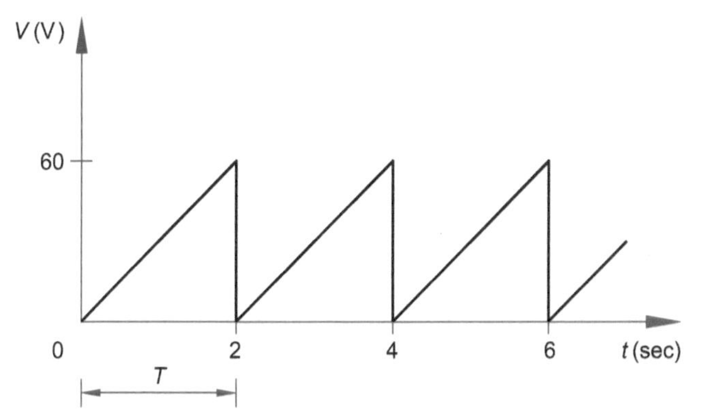

In the waveform, the voltage increases linearly. The root-mean-square (rms) voltage is most nearly

(A) 30 V

(B) 35 V

(C) 64 V

(D) 220 V

140. A DC series motor is supplied with 110 V, and the input current is 20 A. The armature resistance is 0.4 Ω, and the field resistance is 0.2 Ω. The other losses add up to 250 W, and the voltage drop at the brushes is 3 V. The efficiency of the motor is most nearly

(A) 45%

(B) 55%

(C) 65%

(D) 75%

141. Several nonparallel forces hold a rigid body in equilibrium. Which of the following statements is true about the forces?

(A) The forces must be coplanar.

(B) The forces must be concurrent.

(C) The forces must be equal in magnitude.

(D) The forces must form an equilateral triangle.

142. Four truss elements are connected at a node as shown. Force F_1 is 10 kN compressive, and force F_4 is 20 kN compressive.

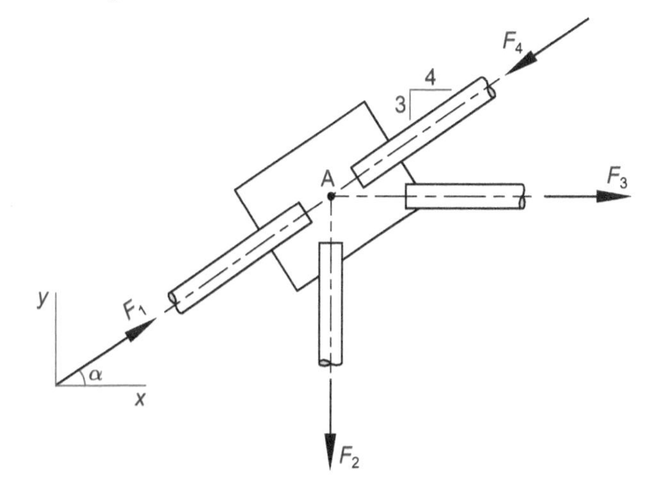

Forces F_2 and F_3 are, respectively,

(A) 6 kN (compression), 8 kN (tension)

(B) 6 kN (tension), 8 kN (compression)

(C) 6 kN (tension), 8 kN (tension)

(D) 6 kN (compression), 8 kN (compression)

143. A shaft is subjected to loading as shown.

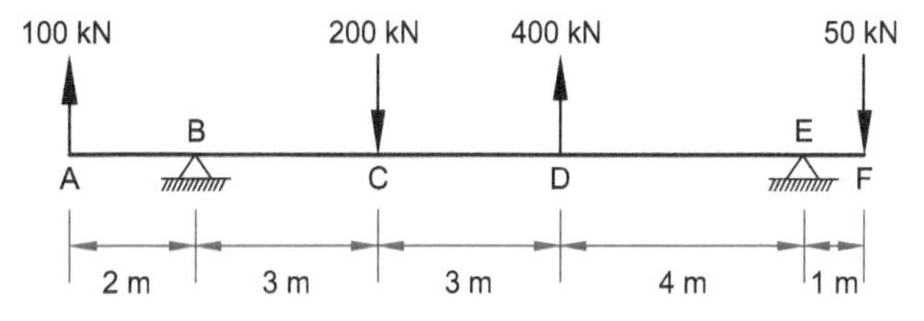

The equivalent loading at point A is most nearly

(A) −250 kN, 1550 kN·m (cw)

(B) 250 kN, 0 kN·m

(C) 250 kN, 1550 kN·m (cw)

(D) 350 kN, 0 kN·m

144. A three-dimensional space truss is subjected to a combination of loads. The truss members may be subjected to

(A) axial force only

(B) axial force and shear force

(C) axial force, shear force, and bending moment

(D) axial force, shear force, bending moment, and torsion

145. A truss tower is shown.

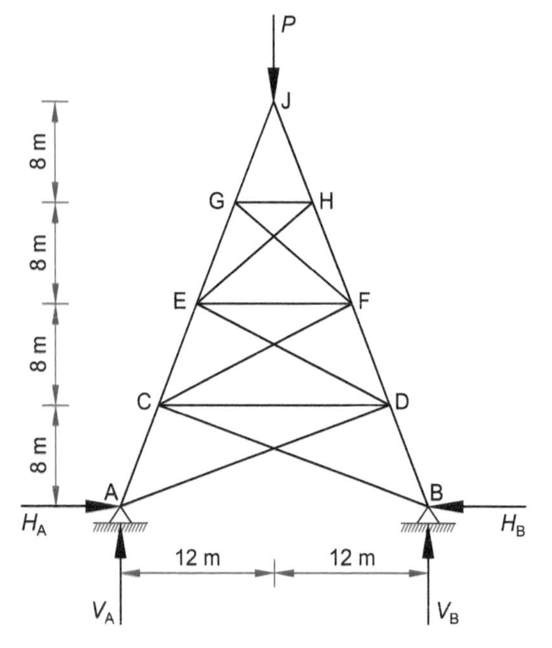

At joint J, the tower carries an antenna that weighs 500 kN. The vertical reaction at support A is most nearly

(A) 25.5 kN

(B) 250 kN

(C) 333 kN

(D) indeterminate, since the tower is indeterminate

146. A composite beam-slab section is shown.

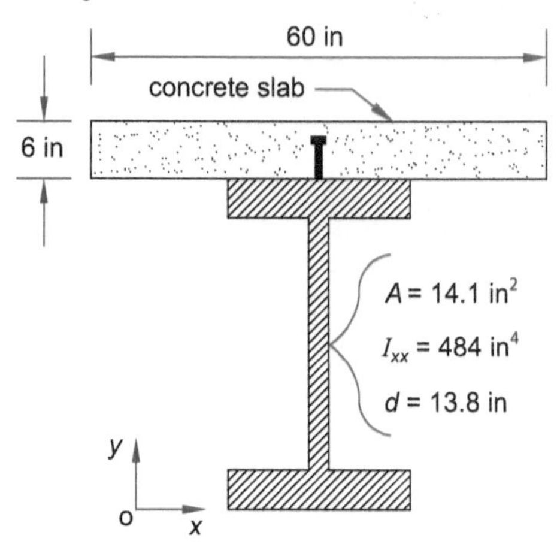

The concrete strength is 5000 psi, and the steel section is a 14 in × 48 in beam ($F_y = 60$ ksi). The distance from the bottom flange to the centroid of the composite section is most nearly

(A) 9.92 in

(B) 10.8 in

(C) 13.8 in

(D) 16.4 in

147. Two identical steel bars are welded together, one on the top of the other. The ratio of the moment of inertia (MOI) of the welded bars to the MOI of a single bar is most nearly

(A) 2

(B) 4

(C) 8

(D) 16

148. A concrete block weighing 150 lbf is resting on a plane surface. Which statements are true? Select **all** that apply.

(A) Limiting friction is defined as the maximum frictional force, which comes into play when a body just begins to slide over the surface of another body.

(B) The force of friction between the two surfaces increases if the contact area is reduced.

(C) The force of friction between the two surfaces decreases if the contact area is reduced.

(D) The force of friction between the two surfaces is independent of the contact area.

(E) The force of friction between the two surfaces is independent of the speed at which the block slides on the surface.

149. A block weighing 500 lbf is placed on a sloping plane with a friction coefficient of 0.5. The block is connected to a cable that is in tension as shown.

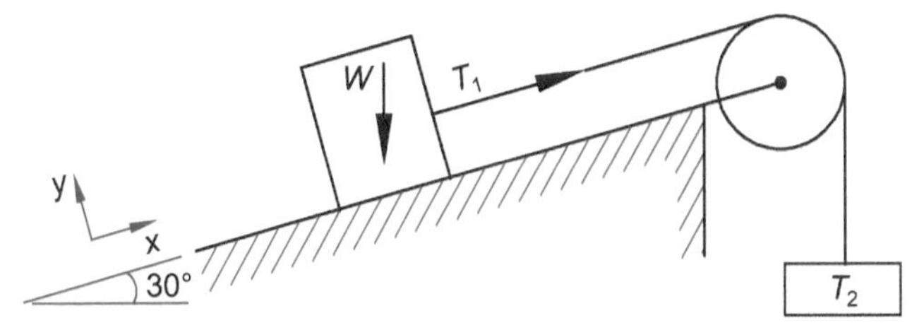

The friction coefficient between the cable and the pulley is 0.2. The minimum force, T_2, required to pull the block up the slope is most nearly

(A) 51 lbf

(B) 110 lbf

(C) 250 lbf

(D) 500 lbf

150. An asymmetrical truss is shown.

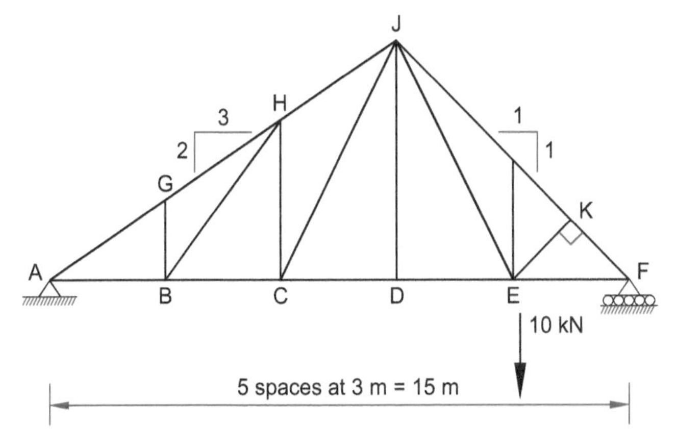

It carries a vertical load of 10 kN at joint E. Which members in the truss have zero force?

(A) AB, EJ, JD, JC

(B) EK, FK, AB, BG

(C) BG, CH, DJ, EJ

(D) EK, DJ, CJ, CH, BH, BG

151. Which statement is true?

(A) Work is said to be done if force is applied and no displacement takes place.

(B) Work is said to be done if no force is applied and displacement takes place.

(C) A body is said to have energy even though it is not moving.

(D) The total energy possessed by a moving body at a constant velocity varies with time.

152. A 10,000 lbf vehicle traveling at 60 mph comes to a stop without skidding. The wheelbase and center of gravity of the vehicle are shown.

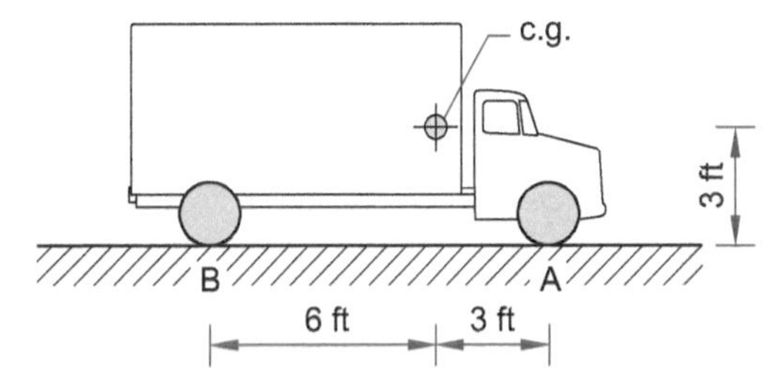

The coefficient of friction is 0.35. The additional load that will be imposed on the front axle due to deceleration is most nearly

(A) 1200 lbf

(B) 3500 lbf

(C) 6500 lbf

(D) 8800 lbf

153. An elevator is designed to carry a 1000 lbf load, including its self-weight. The elevator is hung by a single cable, as shown.

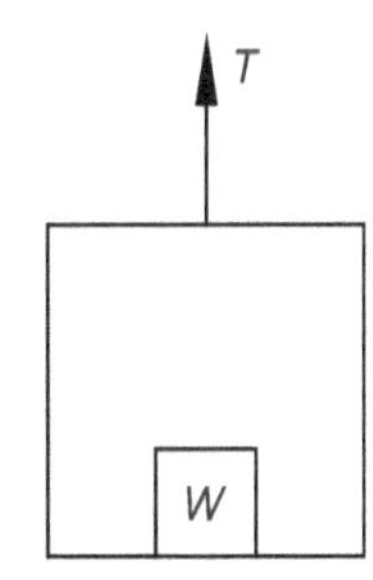

The elevator accelerates upward at a constant rate of 10 ft/sec^2. The service load on the cable is most nearly

(A) 310 lbf

(B) 690 lbf

(C) 1000 lbf

(D) 1300 lbf

154. A 1 kip force is used to compress a spring by 4 in from its free length of 8 in. The spring acts linearly. The work done in compressing the spring an additional 4 in is most nearly

(A) 2 in-kip

(B) 4 in-kip

(C) 6 in-kip

(D) 8 in-kip

155. A crank shaft is 50 cm long and rotates about point A, as shown.

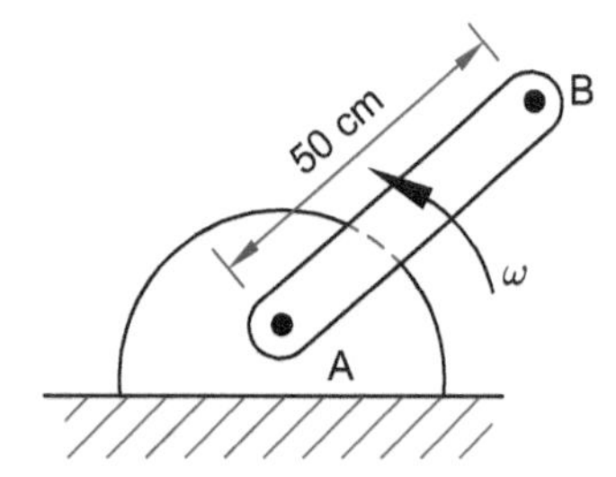

If the shaft is rotating at a constant angular velocity of 20 rad/s, the acceleration at its tip, B, is most nearly

(A) 20 m/s^2

(B) 200 m/s^2

(C) 310 m/s^2

(D) 1000 m/s^2

156. According to Kennedy's rule, when three bodies move relative to one another their instantaneous centers will lie on

(A) a straight line

(B) a circular curve

(C) a parabolic curve

(D) an elliptical curve

157. Which one of the following statements is correct?

(A) The number of centers of gravity of the mass of the body depends on the number of principal axes of the body.

(B) The radius of gyration is defined as the square of the distance from a given reference where the whole mass or area of the body is assumed to be concentrated to give the same value of the moment of inertia.

(C) The maximum moment of inertia is at the centroidal axis of the section.

(D) The least moment of inertia is at the centroidal axis of the section.

158. The Charpy test machine is used to determine the impact load capacity of a test specimen, as shown.

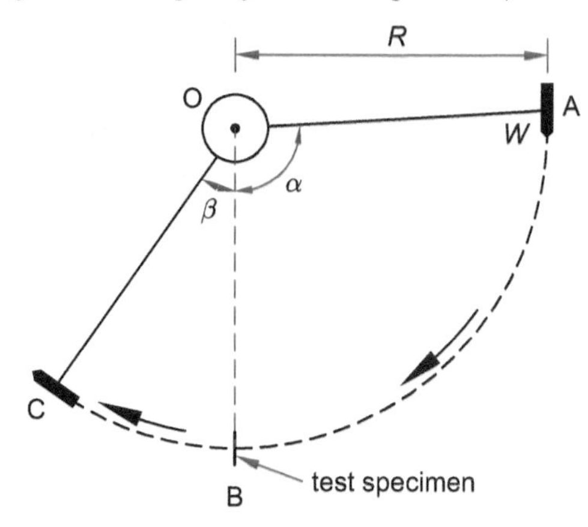

The machine has a 10 lbf striking pendulum. The pendulum's center of mass is at a distance of 3 ft from the pivot point of the pendulum. The scale of the machine is zero degrees in the vertical line and is graduated on both sides from zero. During an impact test, the angle of fall was 90°, and the angle of rise after breaking the specimen was 15°. The energy required to break the specimen is most nearly

(A) 13 ft-lbf

(B) 19 ft-lbf

(C) 29 ft-lbf

(D) 33 ft-lbf

159. A car weighing 6000 lbf and travelling at a speed of 30 mph crashes into a rigid wall, as shown.

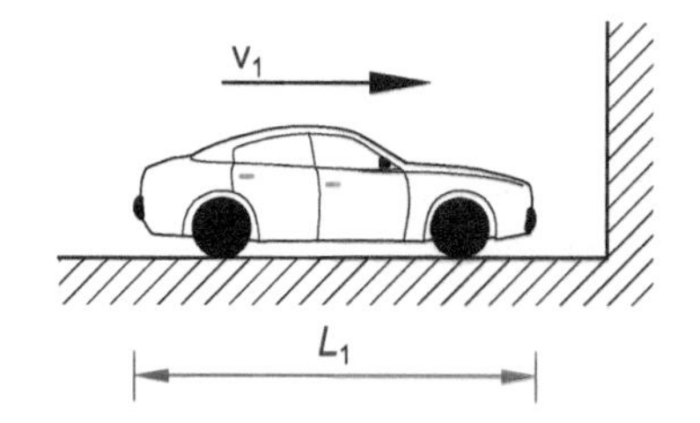

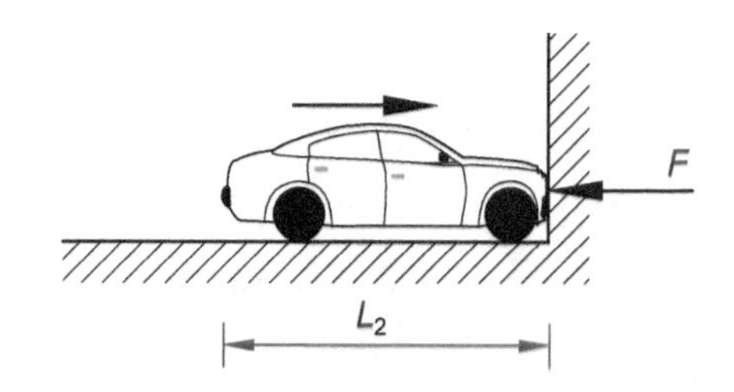

The car front-end deforms instantaneously so that the car length is reduced by 2 ft, as shown. Assume that the car front end deforms plastically, as shown.

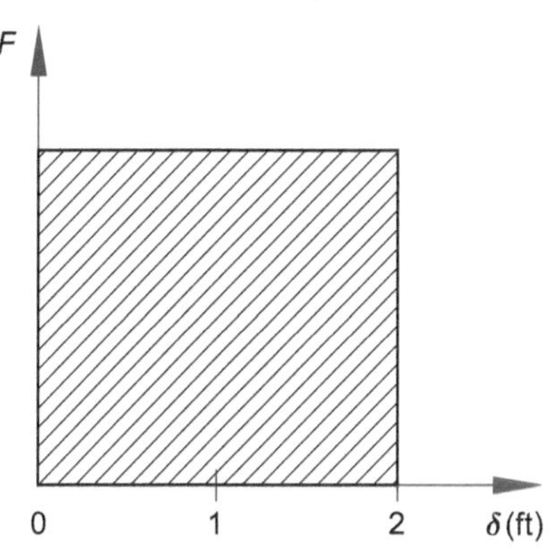

What, most nearly, is the impact force exerted on the car?

(A) 6.1 kips

(B) 62 kips

(C) 91 kips

(D) 134 kips

160. A spring has a stiffness of 8 kips/ft and is carrying a 4 lbf weight. There is no damping device attached to the spring, as shown.

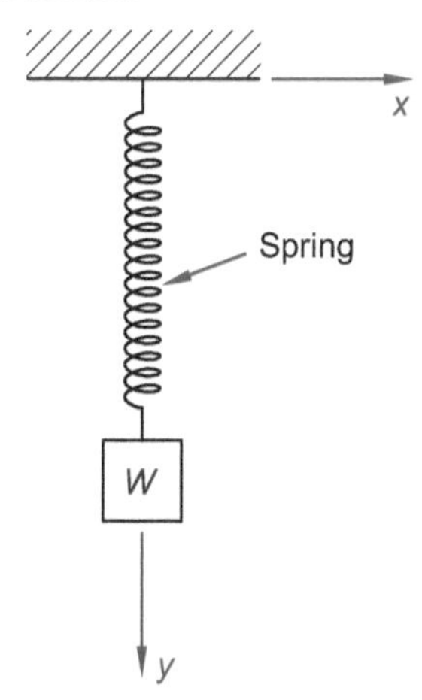

Assume g = 32 ft/sec^2. Which of the following is the equation of motion of the spring?

(A) $\ddot{y} + 64y = 0$

(B) $\ddot{y} + 32y = 0$

(C) $\ddot{y} + 8y = 0$

(D) $\ddot{y} + 4\dot{y} + 32 = 0$

161. Consider a simply supported beam and its shear force (S.F.) diagram shown. The beam loading information is not available.

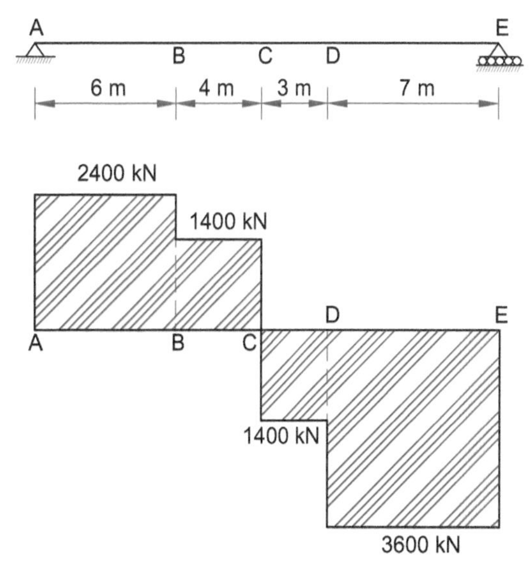

The maximum bending moment in the beam is most nearly

(A) 0 kN·m

(B) 15,000 kN-m

(C) 20,000 kN·m

(D) 40,000 kN·m

162. A point in a thin metallic plate is under biaxial stress of 100 ksi in the x-direction and 50 ksi in the y-direction, as shown.

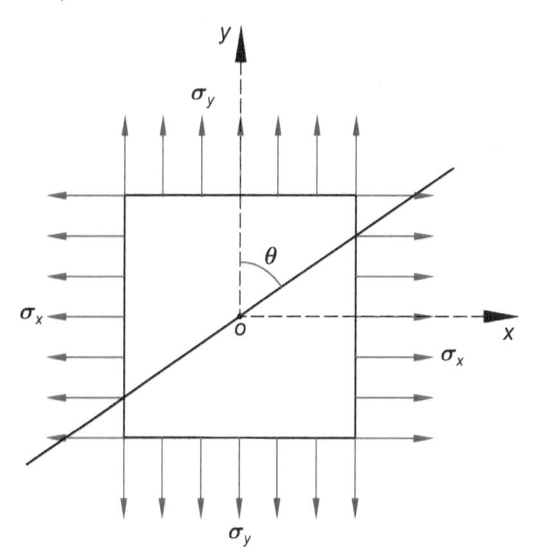

The normal stress at 30° from the minor axis as shown is most nearly

(A) 25 ksi

(B) 35 ksi

(C) 50 ksi

(D) 75 ksi

163. A 1 m long, 20 cm wide, and 20 mm thick aluminum bar is pulled by a 300 kN axial force. The decrease in bar width is most nearly

(A) 1.0 μm

(B) 36 μm

(C) 72 μm

(D) 90 μm

164. A steel bar is placed on top of an identical bar, and then the two are welded together, as shown.

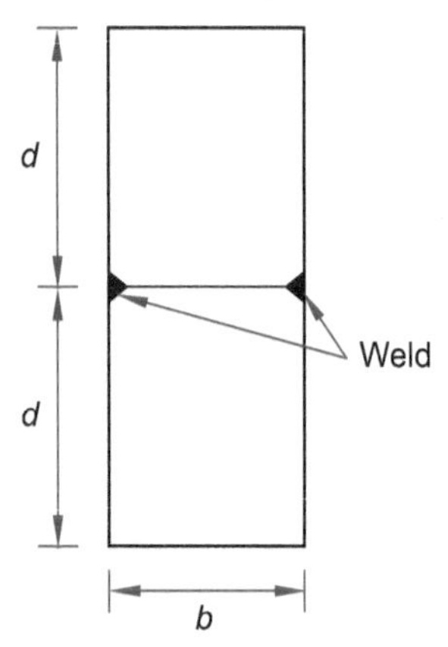

The weld is adequate, and no slippage can take place during bending. What is the ratio of the combined bar's bending stiffness after welding to the bending stiffness of the two bars placed one on top of the other before welding?

(A) 2:1

(B) 4:1

(C) 8:1

(D) 16:1

165. A 12 in outer diameter steel tube has a wall thickness of 1 in. The tube is 10 ft long and is subjected to a torque of 100 ft-kips at its ends. The angle of twist between its ends is most nearly

(A) 0.18°

(B) 0.71°

(C) 1.6°

(D) 3.1°

166. Which of the following is a correct statement?

(A) The elongation of a bar in the direction of the force is called longitudinal strain.

(B) Poisson's ratio is defined as the ratio of longitudinal strain to lateral strain.

(C) The bulk modulus is defined as the ratio of linear stress to linear strain.

(D) The volumetric strain is defined as the ratio of change in volume to original volume.

167. Two bars, A and B, of equal lengths are joined and act compositely. Bar A has higher modulus of thermal expansion than bar B. When the system temperature is raised, what type of stress will be induced in bar A?

(A) compressive stress

(B) tensile stress

(C) shear stress

(D) zero stress

168. A crane runway bracket is subjected to combined shear force and bending moment, as shown. The bracket has four 1 in diameter A325 bolts and a 50 ksi steel plate, which are adequate for the load shown.

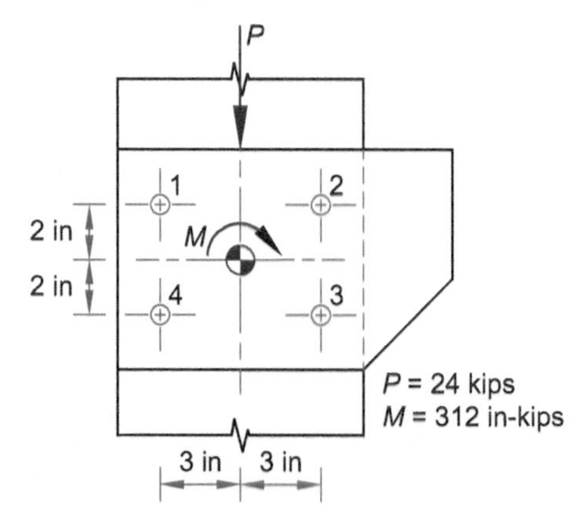

The centroid of the bolt assembly and the location properties of the bolts are tabulated.

no.	x	y	x^2	y^2
1	−3 in	2 in	9 in^2	4 in^2
2	3 in	2 in	9 in^2	4 in^2
3	3 in	−2 in	9 in^2	4 in^2
4	−3 in	−2 in	9 in^2	4 in^2
		sum	36 in^2	16 in^2

The total force in bolt 1 is most nearly

(A) 6.4 kips

(B) 12 kips

(C) 23 kips

(D) 29 kips

169. A steel shaft has an outer diameter of 400 mm and a wall thickness of 20 mm. The shaft is supported at two simple supports 12 m apart. The shaft is carrying a uniformly distributed load (UDL) of 2.5 kN/m, including its self-weight. The shaft's maximum deflection is most nearly

(A) 1.5 mm

(B) 7.8 mm

(C) 16 mm

(D) 48 mm

170. A W14×74 shape grade 50 steel column is shown. The column is 20 ft long and is pinned at both ends.

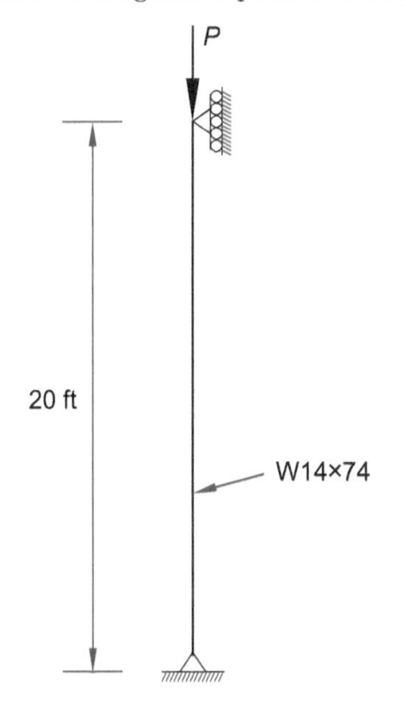

The column's buckling capacity is most nearly

(A) 330 kips

(B) 670 kips

(C) 1100 kips

(D) 1300 kips

171. Consider a prismatic propped cantilever beam. A moment is applied at its non-fixed end. The carryover factor is

(A) zero

(B) one-half

(C) one

(D) two

172. A specimen is tested using the Brinell hardness test and determined to have a BHP number of 98. Its tensile strength is most nearly

(A) 9.8 ksi

(B) 49 ksi

(C) 74 ksi

(D) 98 ksi

173. The stress at which a material will experience permanent deformation is called the

(A) yield stress

(B) ultimate stress

(C) elastic limit

(D) proportional limit

174. An aluminum type Al 2014-T651 plate has a 0.58 in long crack in its center. The plate is subjected to tension as shown.

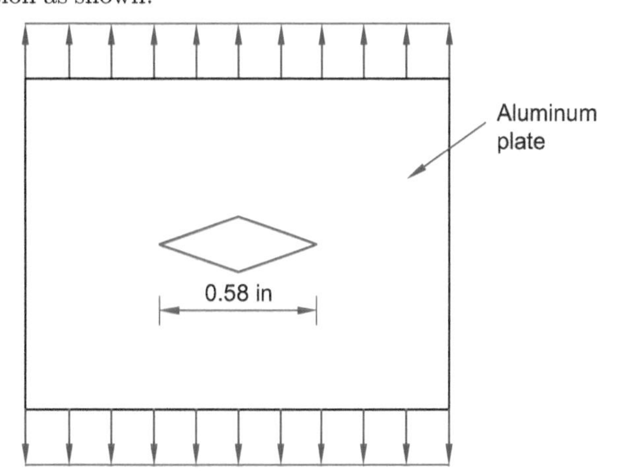

Assume that the plate is infinitely wide and its geometrical factor is unity. The critical value of stress intensity at which catastrophic crack propagation will occur is most nearly

(A) 23 ksi

(B) 29 ksi

(C) 36 ksi

(D) 651 ksi

175. Which of the following statements is correct?

(A) Hardenability is a measure of resistance of plastic deformation as measure by indentation.

(B) Hardenability and hardness are synonymous.

(C) Hardenability can be measured in the state the metal is in.

(D) Hardenability can be gauged by using the Jominy curves.

176. Which one of the following statements is true about cold work?

(A) It improves toughness and impact strength.

(B) It does not affect hardness.

(C) It is measured only qualitatively.

(D) It produces good surface finish on the metal.

177. Which one of the following statements is INCORRECT?

(A) Corrosion is a chemical reaction in which there is transfer of electrons from one chemical species to another.

(B) There is no method of measuring absolute value of an electrode potential.

(C) The metals having electrode potential lower than that of hydrogen are known as anodic or active metals.

(D) In corrosion, surface of the object changes from one element into another element.

178. A hip joint is a ball-and-socket type of joint and is susceptible to fatigue and fracture. The materials that can be considered for the joint replacement are listed in the *NCEES Handbook*, Mechanics of Materials: Average Mechanical Properties of Typical Engineering Materials table. Assuming the materials have comparable biocompatibility and corrosion resistance when used in hip joint replacement, the material that is the most suitable for the joint replacement is ______. Enter your response in the blank.

179. A vehicle rubber tire is inflated so that its pressure gauge reading is 29.5 psi of air under standard atmospheric pressure. Its atmospheric pressure value is most nearly

(A) 0.5 atm

(B) 1.0 atm

(C) 2.0 atm

(D) 5.0 atm

180. A dam is 100 ft long and 13 ft high from the base. The water behind the dam is 12 ft deep. The total lateral force in kips caused by the hydrostatic pressure at the base of the dam, rounded to the nearest integer, is most nearly ______. Enter your response in the blank.

181. A nozzle with a 1 in diameter delivers a jet of 50°F water at a velocity of 300 ft/sec. A perpendicular plate is installed which moves at a speed of 50 ft/sec in the direction of the water jet. The work done by the jet is most nearly

(A) 50 hp

(B) 60 hp

(C) 70 hp

(D) 80 hp

182. A 6 in diameter pipeline is 1000 ft long. Oil weighing 57 pcf and having a kinematic viscosity of 0.02 ft^2/sec is pumped through the pipe. If 22 tons/hr of oil is pumped through the pipeline, the Reynolds number of the flow is most nearly

(A) 31

(B) 2300

(C) 3100

(D) 93,000

183. A billboard 2 m wide and 1 m tall is mounted on top of a 5 m long pole. The pole diameter is 20 cm. The billboard is subjected to 65 km/h wind. The kinematic viscosity of air is 1.47×10^{-5} m^2/s.

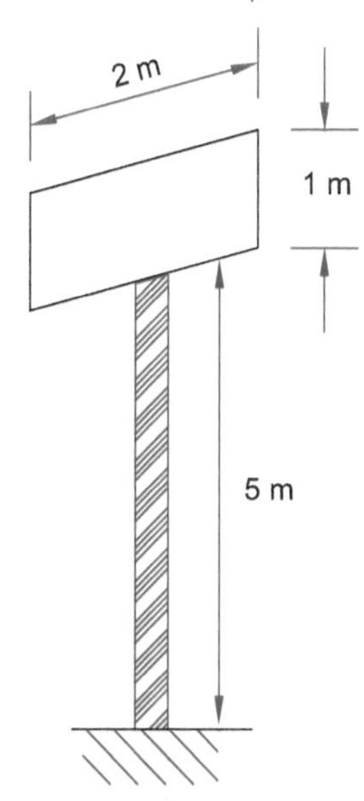

What, most nearly, is the drag coefficient on the pole?

(A) 0.4

(B) 1.0

(C) 2.1

(D) 3.0

184. Dry saturated steam at 10 bar with negligible velocity enters a convergent nozzle and is discharged at a pressure of 1 bar. Assuming no losses, the velocity of steam leaving the nozzle is most nearly

(A) 10 m/s

(B) 103 m/s

(C) 143 m/s

(D) 453 m/s

185. A centrifugal pump runs at a speed of 750 rpm. In order to double the discharge, it is proposed to install two such pumps in parallel for the pumping against a static head of 50 ft through 4000 ft of 12 in main. Both pumps have identical efficiencies. The friction loss of the pipeline for the pumps running in parallel as compared to a single pump operating is most nearly

(A) 0.5

(B) 1

(C) 2

(D) 4

186. A family of pumps has a fixed impeller diameter of 6 3/4 in, an operating speed of 1750 rpm, and a discharge rate of 80 gpm. What, most nearly, is the brake power (BHP) requirement?

(A) 3/4 hp

(B) 1 hp

(C) 1 1/2 hp

(D) 2 hp

187. A 1 hp, 6 in diameter impeller pump with an operating speed of 1500 rpm is being replaced by a pump having the same geometrical design family operating under similar dynamics. The new pump has a diameter of 8 in and an operating speed of speed of 2000 rpm. Both pumps are used for pumping water. The power requirement for new pump is most nearly

(A) 2 hp

(B) 4 hp

(C) 8 hp

(D) 10 hp

188. A dam is modeled on a scale of 16:1. The protype requires a velocity of 4 ft/sec. The velocity at which the model should be tested is

(A) 0.25 ft/sec

(B) 1 ft/sec

(C) 2 ft/sec

(D) 4 ft/sec

189. A spherical vessel is 2 m in diameter and contains air at 19°C and at 750 mm mercury. The air mass in the vessel is most nearly

(A) 0.15 kg

(B) 0.50 kg

(C) 1.1 kg

(D) 5.0 kg

190. A gas mass has a volume of 1 m^3 at a pressure of 2 bar and a temperature of 100°C. It is compressed at a constant pressure to a volume of 0.8 m^3. The work done in compressing the gas is most nearly

(A) 1.2 kJ
(B) 2.5 kJ
(C) 20 kJ
(D) 40 kJ

191. 1 m^3 of propane gas is compressed isentropically from an initial pressure of 1 bar to 15 bar at 289K. The temperature change is most nearly

(A) 16°C
(B) 33°C
(C) 57°C
(D) 81°C

192. A steam turbine is used to generate power. 5 lbm/sec of steam enters the turbine with a velocity of 50 ft/sec and enthalpy of 1000 Btu/lbm. The steam is expanded to a velocity of 100 ft/sec and an enthalpy of 800 Btu/lbm as it enters the condenser. The heat loss from the turbine casing is 49 Btu/sec. Assuming no loss in potential energy, the power generated at the turbine shaft is most nearly

(A) 800 Btu/sec
(B) 950 Btu/sec
(C) 1000 Btu/sec
(D) 1050 Btu/sec

193. A steam power plant is supplied with dry saturated steam at a pressure of 200 psi. It exhausts into a condenser at 2.5 psi and has a saturated liquid enthalpy level of 102 Btu/lbm. The Rankine (isentropic) efficiency of the turbine is most nearly

(A) 26%
(B) 32%
(C) 36%
(D) 63%

194. A diesel engine has a compression ratio of 10:1. Its fuel is cut off at 7% of the stroke. Using the cold air standard, its efficiency is most nearly

(A) 36%
(B) 51%
(C) 56%
(D) 61%

195. HFC-134a refrigerant is used in a refrigeration cycle between 0.1 MPa and 1 MPa. Its coefficient of performance (COP) is most nearly

(A) 0.4
(B) 1.0
(C) 1.6
(D) 2.5

196. A plant produces ice at 10,000 kg daily. The average water supply is at 23.2°C, and the latent heat of ice is 335 kJ/kg. The average heat extracted is most nearly

(A) 50 kW
(B) 135 kW
(C) 335 kW
(D) 450 kW

197. Atmospheric air at 29.921 in of mercury pressure has 80°F dry-bulb temperature and 60°F wet-bulb temperature. Using the psychrometric chart such as given in the *NCEES Handbook*, the corresponding relative humidity, expressed as a percentage and rounded to the nearest integer, is ______. Enter your response in the blank.

198. 4 kg of oxygen is available to burn 1 kg of carbon. The amount of carbon monoxide produced as a result of perfect combustion, in kilograms and rounded to the nearest integer, is ______. Enter your response in the blank.

199. A 160 mm thick composite wall is composed of three different materials.

material type	thickness (mm)	conductivity (W/m·K)
1 outside	100	0.1
2 middle	35	0.07
3 inside	25	0.05

The wall surface is 50 m^2, and the temperature drop from outside to inside is 30°C. The wall's thermal conductance is most nearly

(A) 1 °C/W
(B) 2 °C/W
(C) 4 °C/W
(D) 8 °C/W

200. An air duct has a square section and a diagonal partition as shown.

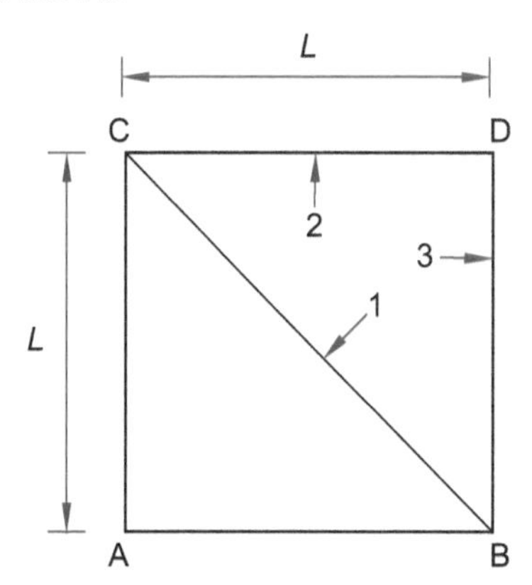

Assuming that the enclosed surfaces are diffuse emitters and reflectors and have uniform radiosity, the shape factor F_{21} of the triangular part is most nearly

(A) 0.31

(B) 0.51

(C) 0.71

(D) 0.91

201. A chilled fluid flows through a long, gray, diffuse tube as shown.

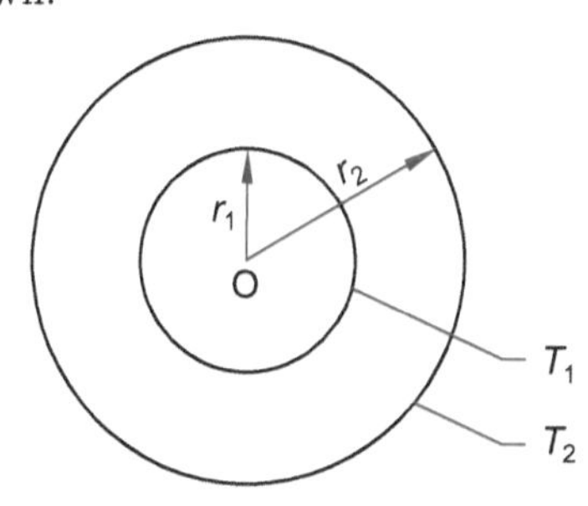

inner tube diameter	20 mm
outer tube diameter	40 mm
temp at outer surface of inner tube	100K
temp at inner surface of outer tube	300K
emissivity for both surfaces, ε	0.2

If the space between the tubes is evacuated, the energy exchange by radiation per unit length from the inner pipe to the outer tube is most nearly:

(A) −6.3 W/m

(B) −4.1 W/m

(C) 3.4 W/m

(D) 6.3 W/m

202. A steel pipe carrying steam is hung from the ceiling in a ventilated room, as shown.

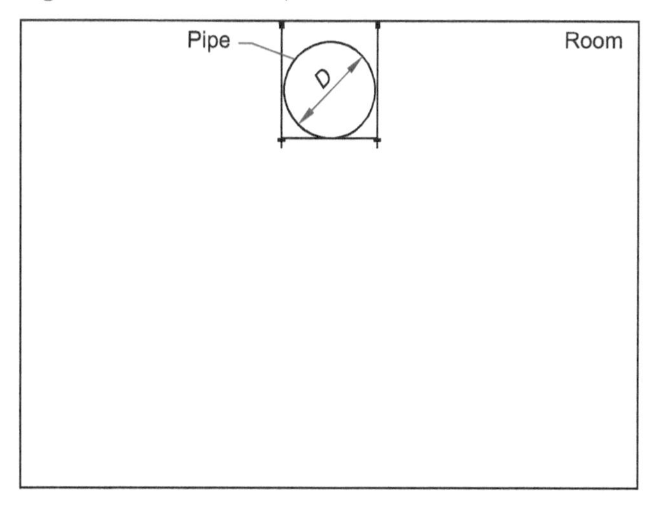

The pertinent heat transfer data is tabulated below.

steel pipe diameter	500 mm
temp at outer surface of pipe	100°C
room (air and walls) steady state temp	27°C
coefficient of heat conduction	83.5 W/m·K
coefficient of convection from pipe to air	15 W/m^2
emissivity for steel pipe outer surface	0.6

The convection and radiation heat loss from pipe surface per meter length is most nearly

(A) 620 W/m

(B) 1220 W/m

(C) 1720 W/m

(D) 2320 W/m

203. An iron pipe, 1 m in diameter, has 35 mm thick walls. The walls are at 0°C and are well insulated. Hot oil is pumped through the pipe. The Fourier number is defined as $\alpha t/L^2$. 5.5 min after the oil starts flowing, the Fourier number at the exterior surface of the pipe is most nearly

(A) 1.5

(B) 6.3

(C) 9.4

(D) 11

204. Hot oil enters a counterflow heat exchanger at 100°C and leaves at 54°C. Water is used to cool the oil in the exchanger. It is specified that water leaves the exchanger at 40°C. The specific heat of the oil is 2 kJ/kg·K. The oil flow rate is 1 kg/s, and the water flow rate is 2 kg/s. Assuming conservation of energy and no losses, the water temperature at the inlet is most nearly

(A) 9.2°C

(B) 19°C

(C) 29°C

(D) 89°C

205. A thermocouple is used to measure temperature in a gas pipe. The thermocouple is spherical in shape with a diameter of 9 mm. Its thermal conductivity is 20 W/m·K. The convection heat-transfer coefficient of the gas is 400 W/m²·K. Its Biot number is ______. Enter your response in the blank.

206. A negative feedback control system model is shown.

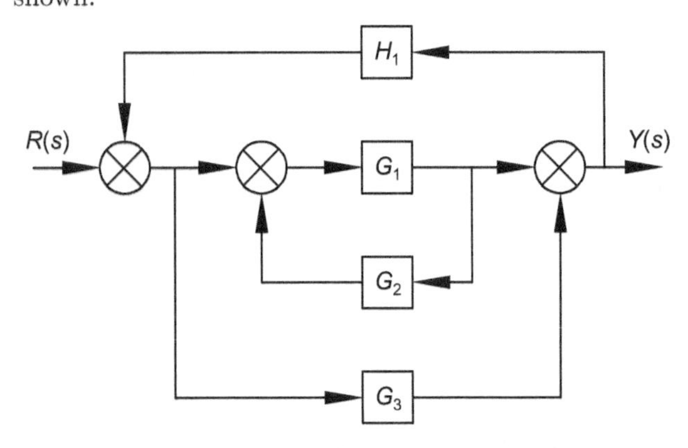

For the reference input $R(s)$, the overall transfer function for the system is expressed as

$$Y(s) = \frac{G_1 + G_3 + G_1G_2G_3}{1 + G_1G_2 - A - B - G_1G_2G_3H_1}$$

The function noted above has two missing terms, A and B. From the following list, move the two correct terms to the equations below.

G_1G_3
G_1H_1
G_2G_3
G_2H_1
G_3H_1

$A =$
$B =$

207. A negative feedback control system model is shown.

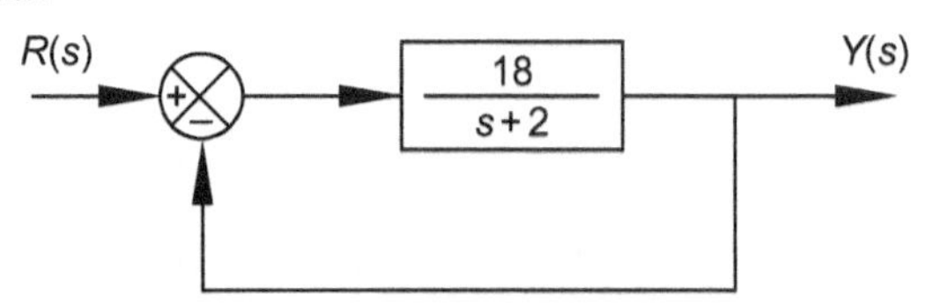

The steady-state gain for the system is most nearly

(A) 0.9

(B) 3

(C) 10

(D) 19

208. A 22.5 kg ± 0.1 kg sphere is moving with a velocity of 5.19 m/s ± 0.05 m/s. The sphere's kinetic energy, including its measurement uncertainty, in joules and rounded to the nearest integer, is ______. Enter your response in the blank.

209. A 350 Ω strain gauge is used to measure axial elongation in a 10 in long metallic bar. The change in resistance is noted as 0.28 Ω. If the gauge factor is 2.0, the elongation in the bar is most nearly

(A) 0.40 μin

(B) 400 μin

(C) 4000 μin

(D) 40,000 μin

210. A typical thermistor is used to check a fluid temperature, and a resistance reading of 9.8 kΩ is noted. The fluid temperature, in degrees Celsius and rounded to the nearest integer, is ______. Enter your response in the blank.

211. A 1 in diameter steel shift linkage is subjected to a tensile force of 100 lbf as shown.

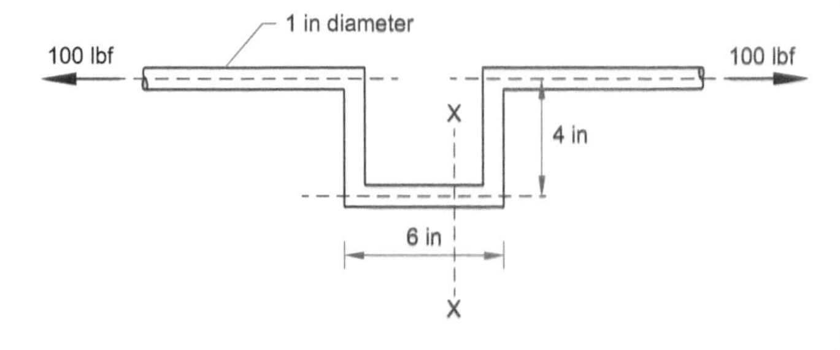

The maximum von Mises stress in plane X-X is most nearly

(A) 1.2 ksi

(B) 2.2 ksi

(C) 3.3 ksi

(D) 4.2 ksi

212. A metallic crankshaft is subjected to an alternating stress of 60 MPa with a mean stress of 60 MPa. Its properties are as follows.

Yield strength = 600 MPa

Ultimate strength = 1200 MPa

Using the Soderberg theory, the fatigue strength of the crankshaft is most nearly

(A) 66 MPa

(B) 96 MPa

(C) 660 MPa

(D) 960 MPa

213. A 3/4 in diameter steel bolt connects the assembly in tension as shown.

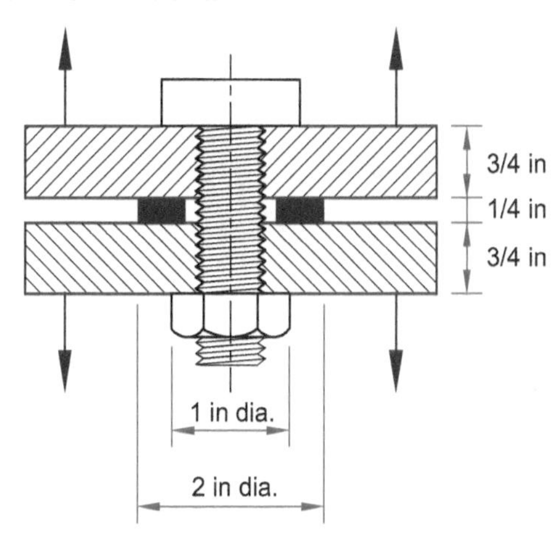

The bolt is properly preloaded, and the plates are assumed rigid. Neglect the difference in bolt stiffness between its threaded and unthreaded portions. The stiffness ratio of the bronze gasket is most nearly

(A) 5.0%

(B) 35%

(C) 65%

(D) 95%

214. Two springs are used to support a load, P, of 100 kg, as shown.

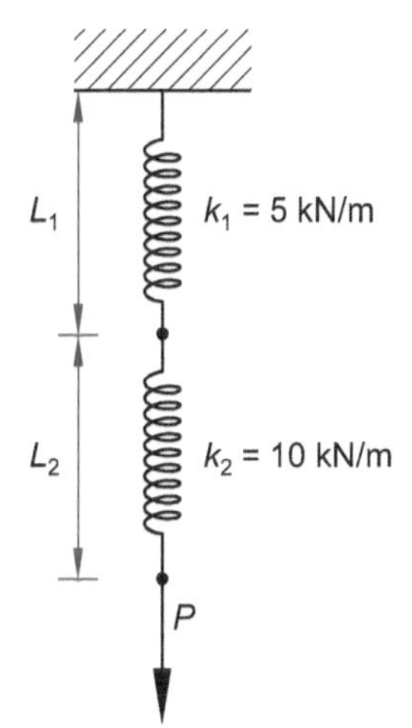

The stiffness of the system is most nearly

(A) 1100 N/m

(B) 3300 N/m

(C) 6700 N/m

(D) 7800 N/m

215. A screw-jack has a single thread of 8 mm pitch on a mean thread radius of 30 mm. The coefficient of friction on the screw threads is 0.12. The magnitude of the torque needed to lower a load of 500 kN, in kN·m and rounded to two decimal places, is ______. Enter your response in the blank.

216. A single disc friction clutch has the following properties.

Outside diameter = 300 mm

Inner diameter = 200 mm

Mean pressure = 0.15 N/mm^2

Coefficient of friction = 0.3

Design speed = 2000 rpm

Assumptions:

1 Both sides of the clutch are equally effective.

2 Torque is uniformly distributed on the ring surface, and its resultant is located at the mean radius of the ring.

The power transmitted by the clutch is most nearly

(A) 9.1 kW

(B) 93 kW

(C) 140 kW

(D) 210 kW

217. A single rivet double cover butt joint in plates 20 mm thick is fastened by 22 mm gross diameter rivets as shown.

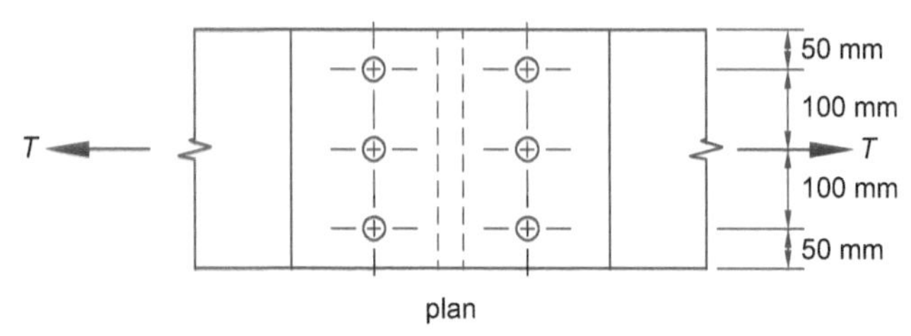

plan

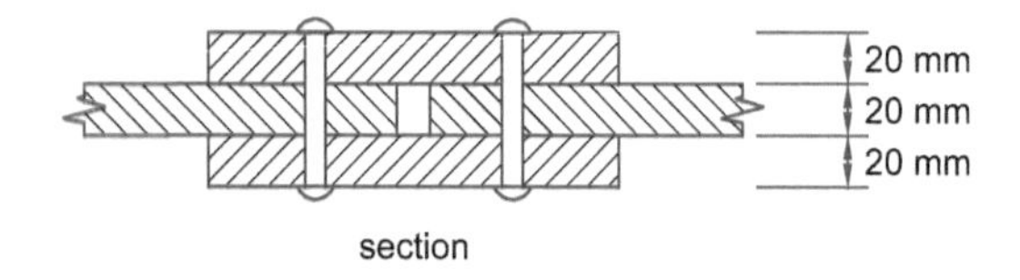

section

The allowable stresses in the rivets are as follows.

Shear stress,

$$\tau = 100 \text{ N/mm}^2$$

Tearing stress,

$$\sigma_t = 150 \text{ N/mm}^2$$

Bearing stress,

$$\sigma_{\text{br}} = 300 \text{ N/mm}^2$$

The joint is in tension, as shown. Its allowable load-carrying capacity is most nearly

(A) 198 kN

(B) 228 kN

(C) 396 kN

(D) 702 kN

218. A manufacturer warrants that their product performs at least 10,000 hours under standard operating conditions. A random sample of 20 units is tested under standard operating conditions until the units failed. Their mean time to failure (MTTF) is tabulated.

no. of units	MTTF (hr)
2	13,000
4	12,000
5	11,000
4	10,000
1	9000
1	8000
1	7000
1	6000
1	5000

Consider that the units are nonrepairable once they fail. The reliability of the product, expressed as a percentage and rounded to one decimal place, is ______. Enter your response in the blank.

219. In preparing mechanical engineering drawings, which is of the following drawing practices is correct?

(A) Specify sheet metal by its gage number.

(B) Specify the method of manufacturing for the component to be manufactured.

(C) Specify all angles between lines.

(D) Specify each necessary dimension of an end product. No more dimensions than those necessary for complete definition shall be given.

220. A cylindrical hole has a specified diameter of 1.00 $\pm$ 0.01. The nominal dimension of the cylinder so that it fits into the hole with clearance of $\pm$ 0.01 is most nearly ____. Enter your response in the blank.

STOP!

DO NOT CONTINUE!

This concludes the examination. If you finish early, check your work and make sure that you have followed all instructions. After checking your answers, you may turn in your examination booklet and answer sheet and leave the examination room. Once you leave, you will not be permitted to return to work or change your answers.

Exam 1 Answer Key

1.	A
2.	D
3.	A
4.	C
5.	A
6.	C
7.	C
8.	C
9.	D
10.	B
11.	C
12.	A
13.	C
14.	A, B, C
15.	D
16.	D
17.	A
18.	C
19.	C
20.	D
21.	A
22.	A
23.	B
24.	B
25.	C
26.	B
27.	A
28.	B, D
29.	C
30.	D
31.	D
32.	C
33.	B
34.	B
35.	C
36.	C
37.	D
38.	A, C
39.	D
40.	B
41.	C
42.	B
43.	B
44.	C
45.	D
46.	B
47.	D
48.	B
49.	C
50.	D
51.	B
52.	B
53.	C
54.	C
55.	A
56.	D
57.	C
58.	C
59.	C
60.	D
61.	B
62.	C
63.	B
64.	C
65.	D
66.	A
67.	C
68.	D
69.	C
70.	B
71.	B
72.	C
73.	B
74.	D
75.	B
76.	See Sol. 76.
77.	C
78.	D
79.	A
80.	A
81.	See Sol. 81.
82.	See Sol. 82.
83.	See Sol. 83.
84.	See Sol. 84.
85.	See Sol. 85.
86.	See Sol. 86.
87.	See Sol. 87.
88.	See Sol. 88.
89.	D
90.	See Sol. 90.
91.	See Sol. 91.
92.	B
93.	B
94.	A
95.	B
96.	D
97.	B
98.	B
99.	See Sol. 99.
100.	C
101.	B
102.	A
103.	B
104.	D
105.	A
106.	C
107.	B
108.	B
109.	B
110.	B

Exam 2 Answer Key

111. (A) (B) ● (D)
112. (A) ● (C) (D)
113. ● (B) (C) (D)
114. (A) (B) ● (D)
115. (A) (B) ● (D)
116. (A) ● (C) (D)
117. ● (B) (C) (D)
118. ● (B) (C) (D)
119. ● (B) (C) (D)
120. (A) ● (C) (D)
121. (A) (B) (C) ●
122. ● (B) (C) (D)
123. (A) (B) ● (D)
124. ● ● ● (D) (E)
125. ● ● ● (D) (E)
126. ● (B) (C) (D)
127. (A) (B) (C) ●
128. (A) ● (C) (D)
129. (A) (B) (C) ●
130. (A) ● (C) (D)
131. (A) (B) ● (D)
132. (A) ● (C) (D)
133. ● (B) (C) (D)
134. ● (B) (C) (D)
135. ● (B) (C) (D)
136. (A) ● (C) (D)
137. (A) ● (C) (D)
138. See Sol. 138.
139. (A) ● (C) (D)
140. (A) (B) (C) ●
141. (A) ● (C) (D)
142. ● (B) (C) (D)
143. (A) (B) ● (D)
144. ● (B) (C) (D)
145. (A) ● (C) (D)
146. (A) (B) (C) ●
147. (A) (B) ● (D)
148. ● (B) (C) ● (E)
149. ● (B) (C) (D)
150. (A) (B) (C) ●
151. (A) (B) ● (D)
152. ● (B) (C) (D)
153. (A) (B) (C) ●
154. (A) (B) ● (D)
155. (A) ● (C) (D)
156. ● (B) (C) (D)
157. (A) (B) (C) ●
158. (A) (B) ● (D)
159. (A) (B) ● (D)
160. ● (B) (C) (D)
161. (A) (B) ● (D)
162. ● (B) (C) (D)
163. (A) (B) ● (D)
164. (A) ● (C) (D)
165. (A) ● (C) (D)
166. (A) (B) (C) ●
167. ● (B) (C) (D)
168. (A) ● (C) (D)
169. (A) ● (C) (D)
170. (A) ● (C) (D)
171. (A) ● (C) (D)
172. (A) ● (C) (D)
173. ● (B) (C) (D)
174. ● (B) (C) (D)
175. (A) (B) (C) ●
176. (A) (B) (C) ●
177. (A) (B) (C) ●
178. See Sol. 178.
179. (A) (B) ● (D)
180. See Sol. 180.
181. (A) ● (C) (D)
182. ● (B) (C) (D)
183. (A) ● (C) (D)
184. (A) (B) (C) ●
185. (A) (B) (C) ●
186. (A) ● (C) (D)
187. (A) (B) (C) ●
188. (A) ● (C) (D)
189. (A) (B) (C) ●
190. (A) (B) (C) ●
191. (A) (B) (C) ●
192. (A) ● (C) (D)
193. ● (B) (C) (D)
194. (A) (B) ● (D)
195. (A) (B) (C) ●
196. ● (B) (C) (D)
197. See Sol. 197.
198. See Sol. 198.
199. (A) ● (C) (D)
200. (A) (B) ● (D)
201. (A) ● (C) (D)
202. (A) (B) (C) ●
203. (A) ● (C) (D)
204. (A) (B) ● (D)
205. See Sol. 205.
206. See Sol. 206.
207. ● (B) (C) (D)
208. See Sol. 208.
209. (A) (B) ● (D)
210. See Sol. 210.
211. (A) (B) (C) ●
212. ● (B) (C) (D)
213. (A) (B) (C) ●
214. (A) ● (C) (D)
215. See Sol. 215.
216. (A) ● (C) (D)
217. (A) ● (C) (D)
218. See Sol. 218.
219. (A) (B) (C) ●
220. See Sol. 220.

Solutions

Exam 1

1. See the *NCEES Handbook*, Mathematics section. The general equation of a straight line is

$$y - y_1 = m(x - x_1)$$

The slope, m, of a line passing through points (x_1, y_1) and (x_2, y_2) is

$$m = \frac{y_2 - y_1}{x_2 - x_1}$$

For the two given points, the slope is

$$\begin{aligned} m &= \frac{y_2 - y_1}{x_2 - x_1} = \frac{0-2}{5-1} = \frac{-2}{4} \\ &= -\frac{1}{2} \end{aligned}$$

It can be represented graphically as

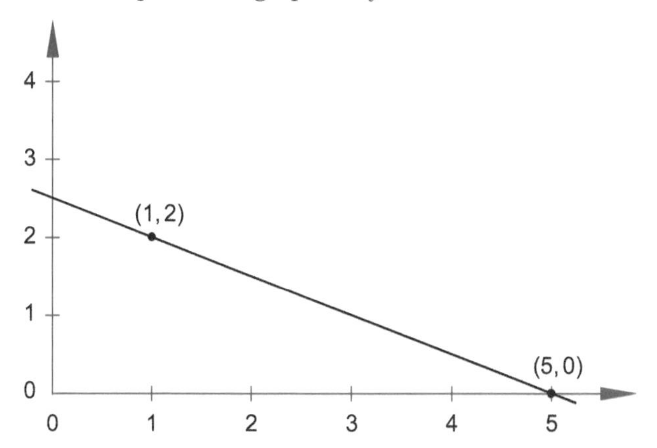

Use the slope and one of the original points to calculate the standard form of the line.

$$\begin{aligned} y - y_1 &= m(x - x_1) \\ y &= mx - mx_1 + y_1 \\ y - mx &= -mx_1 + y_1 \\ y - \left(-\frac{1}{2}\right)(x) &= -\left(-\frac{1}{2}\right)(1) + 2 \\ y + \frac{x}{2} &= \frac{5}{2} \\ 2y + x &= 5 \end{aligned}$$

The standard form of the line passing through the points $(1, 2)$ and $(5, 0)$ is

$$2y + x = 5$$

The answer is (A).

2. The line represented by the equation is shown.

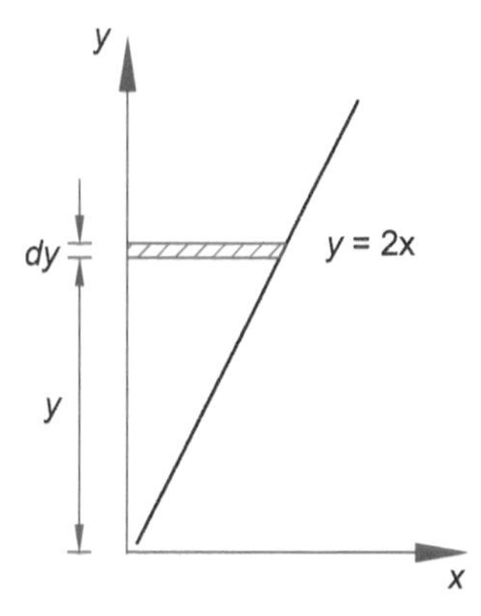

The rotation of the line about its y-axis generates a conical shape. The area of the circle at the top of the conical shape at a given x is

$$A = \pi x^2$$

The volume of a slice of thickness dy is

$$\begin{aligned} dV &= \pi x^2 dy = \pi\left(\frac{y}{2}\right)^2 dy \\ &= \frac{\pi y^2}{4} dy \end{aligned}$$

The volume between $y = 0$ and $y = 10$ is

$$\begin{aligned} V &= \int dV = \int_0^{10} \frac{\pi y^2}{4} dy \\ &= \left(\frac{\pi}{4}\right)\left(\frac{y^3}{3}\right)\Bigg|_0^{10} \\ &= \left(\frac{\pi}{12}\right)(10^3 - 0^3) \\ &= 250\pi/3 \end{aligned}$$

Shortcut: The solution can be arrived at by using the cone volume formula given in the Mensuration of Areas and Volumes section of the *NCEES Handbook*.

The answer is (D).

3. The equation is a first-order linear homogeneous differential equation with constant real coefficients. For the form $y' + ay = 0$, the general solution is $y = Ce^{-at}$. Here, $a = 3$, so $y = Ce^{-3t}$.

The constant C must satisfy the boundary condition. Substitute $y(0) = 1$ into the general solution.

$$y(0) = Ce^{-3(0)} = C$$

Therefore, $C = 1$, and $y = e^{-3t}$.

The answer is (A).

4. See the *NCEES Handbook*, Mathematics section. The parallelogram can be drawn as

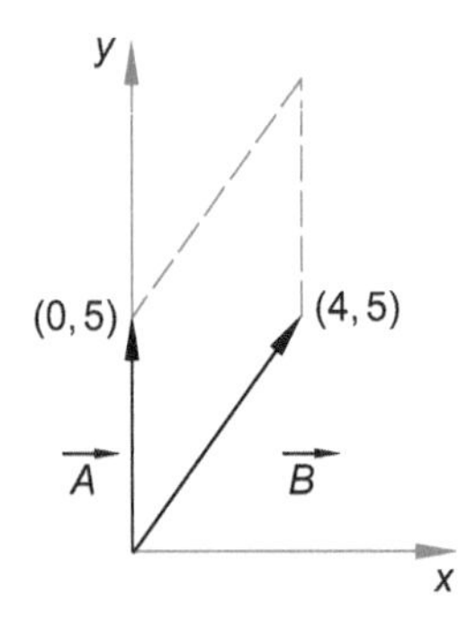

The two vectors are

$$\vec{A} = 0i + 5j$$
$$\vec{B} = 4i + 5j$$

The rule: The magnitude of determinant is the area of parallelogram, with vectors arranged in columns. For two given vectors, the area of the parallelogram is

$$\begin{aligned} \text{Area} &= \left|\det\begin{pmatrix} 0 & 4 \\ 5 & 5 \end{pmatrix}\right| = |(0)(5) - (4)(5)| \\ &= |0 - 20| \\ &= 20 \quad (20.0) \end{aligned}$$

The answer is (C).

5. See the *NCEES Handbook*, Mathematics section, for definition of a Taylor series. In a Taylor series, the function $f(x)$ is expanded about the point a and expressed as

$$f(x) = \begin{pmatrix} f(a) + \left(\dfrac{f'(a)}{1!}\right)(x-a) \\ + \left(\dfrac{f''(a)}{2!}\right)(x-a)^2 + \cdots \\ + \left(\dfrac{f^{(n)}(a)}{n!}\right)(x-a)^n + \cdots \end{pmatrix}$$

If the value of a is zero, then the Taylor's series expansion is called a Maclaurin's series and is expressed as

$$f(x) = \begin{pmatrix} \left(\dfrac{f'(a)}{1!}\right) + \left(\dfrac{f''(a)^2}{2!}\right) \\ + \cdots + \left(\dfrac{f^{(n)}(a)^n}{n!}\right) + \cdots \end{pmatrix}$$

The angle is

$$30° = \frac{(\pi \text{ rad})(30°)}{180°} = \frac{\pi}{6} \text{ rad}$$

The equation in the problem statement is an alternating series of Maclaurin's equation. For alternating series, the rule is that the error in approximating the function cannot be more than the first neglected term in the series. In this problem, the neglected term is $-x^7/7!$.

Determine the maximum error as

$$\begin{aligned} \text{maximum error} &= \left|\frac{x^7}{7!}\right| = \frac{|x^7|}{5040} = \frac{\left(\dfrac{\pi}{6}\right)^7}{5040} \\ &= 2.14 \times 10^{-6} \quad (2 \times 10^{-6}) \end{aligned}$$

The answer is (A).

6. The values of C at each iteration level are shown.

ITER	C = 0.5*(B + A/B)
1	= 0.5(1 + 100/1) = 50
2	= 0.5(50 + 100/50) = 26
3	= 0.5(26 + 100/26) = 14.92
4	= 0.5(14.92 + 100/14.92) = 10.81
5	= 0.5(10.81 + 100/10.81) = 10.03
6	= 0.5(10.03 + 100/10.03) = 10.00

The algorithm is used to determine the square root of any number without using a square root function. The square root of 100 is 10, which is reached in six iterations. The problem asks for the value of C when at ITER = 4, which is 10.81.

The answer is (C).

7. Expansion by minors is used to find the value of a determinant. Using the *NCEES Handbook* notations, calculate the value of a third-order determinant by expanding the determinant.

$$\begin{vmatrix} a_1 & a_2 & a_3 \\ b_1 & b_2 & b_3 \\ c_1 & c_2 & c_3 \end{vmatrix} = a_1\begin{vmatrix} b_2 & b_3 \\ c_2 & c_3 \end{vmatrix} - a_2\begin{vmatrix} b_1 & b_3 \\ c_1 & c_3 \end{vmatrix} + a_3\begin{vmatrix} b_1 & b_2 \\ c_1 & c_2 \end{vmatrix}$$

$$\begin{aligned} &= a_1b_2c_3 + a_2b_3c_1 + a_3b_1c_2 \\ &\quad - a_3b_2c_1 - a_2b_1c_3 - a_1b_3c_2 \end{aligned}$$

Use the above equation to calculate the determinant of the given matrix.

$$\begin{vmatrix} 1 & 2 & -1 \\ 2 & 3 & 2 \\ 1 & -2 & -2 \end{vmatrix} = 1\begin{vmatrix} 3 & 2 \\ -2 & -2 \end{vmatrix} - 2\begin{vmatrix} 2 & 2 \\ 1 & -2 \end{vmatrix} - 1\begin{vmatrix} 2 & 3 \\ 1 & -2 \end{vmatrix}$$

$$\begin{aligned} &= 1(3)(-2) + 2(2)(1) \\ &\quad +(-1)(2)(-2) - (-1)(3)(1) \\ &\quad -2(2)(-2) - 1(2)(-2) \\ &= 1[3(-2) - (2)(-2)] \\ &\quad - 2[2(-2) - 2(1)] \\ &\quad - 1[2(-2) - 3(1)] \\ &= 17 \end{aligned}$$

The answer is (C).

8. As the value of x approaches 27, both the numerator and denominator become zero, which yields the 0/0 form of an indeterminate answer. Use L'Hopital's rule and differentiate the expression.

Use the rule

$$\frac{d(u^n)}{dx} = nu^{n-1}\,dx$$

The first differentiation of the expression is

$$\lim_{x\to 27}\frac{\left(\frac{1}{3}x^{-2/3}\right) - 0}{1-0} = \frac{1}{3(27)^{2/3}} = \frac{1}{(3)(9)} = \frac{1}{27}$$

The answer is (C).

9. In this problem, the order matters. For example, if the required combination to the safe is 123, then 321, 213, and so on, would not work. There are 10 numbers (0 to 9) to select from for each required number. For $n = 10$ and $r = 1$, use the formula for the number of different permutations.

$$_nP_r = \frac{n!}{(n-r)!}$$

$$_{10}P_1 = \frac{10!}{(10-1)!} = 10$$

The process is repeated three times (one for each entry number).

Total permutations = (10)(10)(10) = 1000.

The answer is (D).

10. Arrange the data in either descending or ascending order. The set of data is comprised of 100 workers. Therefore, $n = 100$. The data in ascending order of wages is shown:

No.	No. of workers	Wage ($/hr)	Cumulative frequency	Notes
1	10	$20.00	10	
2	20	$22.00	10 + 20 = 30	
3	20	$25.00	30 + 20 = 50	50th ordered value
4	25	$30.00	50 + 25 = 75	51st ordered value
5	15	$35.00	75 + 15 = 90	
6	10	$40.00	90 + 10 = 100	Total workers, $n = 100$

Since n is an even number, consider the two middle values: the 50th value is $25, and the 51st value is $30. The median is the average.

$$\begin{aligned} \text{Median} &= \frac{25+30}{2} \\ &= \$27.50/\text{hr} \end{aligned}$$

The answer is (B).

11. Determine the expected cost based on the first report.

$$\begin{aligned} \text{cost}_{\text{report 1}} &= \left(500{,}000\ \text{ft}^2\right)\left(0.25\left(\frac{\$100}{\text{ft}^2}\right) + 0.75\left(\frac{\$300}{\text{ft}^2}\right)\right) \\ &= \$125{,}000{,}000 \end{aligned}$$

Determine the expected cost based on the second report.

$$\text{cost}_{\text{report 2}} = (500{,}000\ \text{ft}^2)\left(0.55\left(\frac{\$100}{\text{ft}^2}\right)+0.45\left(\frac{\$300}{\text{ft}}\right)\right)$$
$$= \$95{,}000{,}000$$

Determine the expected cost: Use the equation for the arithmetic mean of a set of values from the Engineering Probability and Statistics section of the *NCEES Handbook*:

$$\begin{aligned}\bar{x} &= \frac{1}{n}\sum_{i=1}^{n} x_i \\ &= \frac{\text{cost}_{\text{report 1}}+\text{cost}_{\text{report 2}}}{2} \\ &= \frac{\$125{,}000{,}000+\$95{,}000{,}000}{2} \\ &= \$110{,}000{,}000 \quad (\$110\ \text{million})\end{aligned}$$

The answer is (C).

12. Use the formula from the *NCEES Handbook*. To evaluate the parameters in the regression equations, tabulate the data.

	x	y	xy	x^2
	2	9	18	4
	3	11	33	9
	5	15	75	25
	9	22	198	81
Σ	**19**	**57**	**324**	**119**

Determine the summed values S_{xy} and S_{xx}, slope $\hat{b}$, and the y-intercept $\hat{a}$.

$$\begin{aligned}S_{xy} &= \sum_{i=1}^{n} x_i y_i - \left(\frac{1}{n}\right)\left(\sum x_i\right)\left(\sum y_i\right) \\ &= 324 - \frac{(19)(57)}{4} = 53.25\end{aligned}$$

$$\begin{aligned}S_{xx} &= \sum_{i=1}^{n} (x_i)^2 - \left(\frac{1}{n}\right)\left(\sum x_i\right)^2 \\ &= 119 - \frac{(19)^2}{4} = 28.75\end{aligned}$$

$$\hat{b} = \frac{S_{xy}}{S_{xx}} = \frac{53.25}{28.75} = 1.85$$

$$\hat{a} = \bar{y} - \hat{b}\bar{x} = \frac{\sum y}{4} - 1.85\left(\frac{\sum x}{4}\right)$$

Use the calculated values from the table.

$$\hat{a} = \frac{57}{4} - 1.85\left(\frac{19}{4}\right) = 5.46$$

Determine the line equation.

$$\begin{aligned}y &= \hat{a} + \hat{b}x \\ &= 5.46 + 1.85x\end{aligned}$$

The answer is (A).

13. Given that the AEP $= 1\% = 0.01$,

$$\begin{aligned}P(\text{no exceedance in one year}) &= 1 - 0.01 = 0.99 \\ P(\text{no exceedance in } n \text{ years}) &= (0.99)^n \\ P(\text{exceedance in } n \text{ years}) &= 1 - (0.99)^{100} \\ &= 1 - 0.366 \\ &= 0.634 \\ &= 63.4\% \quad (63\%)\end{aligned}$$

The answer is (C).

14. Ethics are a set of guidelines, rules, philosophical concepts, customs, norms, and aspirations for a licensee to follow. The Code of Ethics for Engineers articulates the ways in which moral and ethical principles apply to unique situations encountered in professional practice. It indicates to others that the professionals are seriously concerned about responsible and professional conduct.

In some cases, it is impossible to comply with every aspect of the Code. Therefore, ethics are also called a set of aspirations that a licensed engineer should aim for.

However, ethics are not subject to the law. As the late Chief Justice of the U.S. Supreme Court Earl Warren put it, "Society would come to grief without ethics, which is unenforceable in the courts and cannot be made part of law ... Not only does law in a civilized society presuppose ethical commitment, it presupposes the existence of a broad area of human conduct controlled only by ethical norms and not subject to law at all." For this reason, a violation of ethics does not trigger a criminal penalty unless such act is deemed a crime under a criminal statute. Statements D and E are incorrect.

The answer is (A), (B), and (C).

15. See the *NCEES Handbook* section Ethics and Professional Practice, which provides a narration of the model rules. According to the *Model Rules,* Sec. 240.15 (A)(1): "Licensees shall be cognizant that their first and foremost responsibility is to safeguard the health, safety, and welfare of the public when performing services for clients and employers." Therefore, a licensee's first and foremost responsibility in performance of professional services is to the public welfare.

The answer is (D).

16. See the *NCEES Handbook*, Ethics and Professional Practice section. A *copyright* covers original works of authorship, such as a book, song, or movie. Therefore, statement A is correct.

A *trademark* protects words, phrases, symbols, or designs made to identify a distinguishable source of a good or service. Therefore, statement B is correct. A *patent* protects unique inventions and discoveries. Therefore, statement C is correct.

A trade secret applies to a formula, pattern, device, method, technique, process, and so on. To meet the most common definition of a trade secret, it must be used in business and give an opportunity to obtain an economic advantage over competitors who do not know or use it. Trade secrets offer little protection without a written agreement between the involved parties to keep it secret from others. In this case, the inventor intends to use it in business, but has not started using it. Therefore, the secret idea is not recognized as an intellectual property.

The answer is (D).

17. The *NCEES Handbook* does not provide a prescriptive definition of sustainability. It provides those sustainable principles that include a consideration of

- safety
- public health
- quality of life
- resource allocation
- nonrenewable resources

Several definitions of sustainability are available in the engineering world, and every definition has several elements, similar in nature but expressed in different words. In general, a sustainable construction requires that recyclable resources should be used to minimize consumption of virgin resources, and it requires protecting nature. A focus on quality is needed for every successful project.

The answer is (A).

18. See the *NCEES Handbook*, Ethics and Professional Practice: Societal Considerations section. The parameters such as air and water pollution and atmospheric emissions are required in a life-cycle analysis. The *NCEES Handbook* does not explicitly provide information on greenhouse gases. The subject is covered in many textbooks. Greenhouse gases include carbon dioxide and other carbon compounds. There is no standard global method available for calculating the carbon footprint. Comparing the carbon footprint of different companies or products from different producers is extremely difficult. However, emissions can be grouped into categories. For example, the Berkeley Institute of the Environment focuses on transportation, housing, food, and goods and services.

The answer is (C).

19. Option A is incorrect because a standard of perfection cannot be met.

Option B is incorrect because meeting the minimum code requirement may not result in meeting the standard of care. Generally, the code provisions provide the minimum design requirements a design professional must adhere to. Sometimes the minimum requirements are adequate, and sometimes they are not. In other occasions, the code provisions may not address a situation at all. In such situations, local engineering associations or specialty institutes provide the state-of-the-art information for the design issue. For example, an engineer uses the code yet designs a building system that performs poorly in an earthquake. Should the engineer use the same code provision again for the redesign? If not, what design criteria should be used in such a situation? A reasonable approach would be to look beyond the code provisions and incorporate state-of-the-art methods in design.

The engineer is not responsible for job site safety unless the engineer assumes such responsibility by contract or conduct. Therefore, option D is incorrect.

The answer is (C).

20. The following information is given.

$$\begin{aligned} P &= \$24 \\ i &= 7\% = 0.07 \\ n &= 2016 - 1626 = 390 \text{ yr} \end{aligned}$$

Determine the future value given the present value by using the formula provided in the Engineering Economics section of the *NCEES Handbook.*

$$\begin{aligned} (F/P,\ i\%,\ n) &= (1+i)^n \\ F &= P(1+0.07)^{390} \\ &= \$24(2.882\times 10^{11}) \\ &= \$6.912\times 10^{12} \quad (\$7 \text{ trillion}) \end{aligned}$$

The answer is (D).

21. The problem presents two mutually exclusive alternatives, out of which one alternative must be selected. There is no "do nothing" choice. The income and expenses for each option are converted to the equivalent present values, also called the *P*-pattern. The given annual interest rate is 8%. The *P*-value is the present net equivalent value of the stream of all future incomes and expenses. The alternative with the lowest *P*-value is selected.

The *P*-pattern values of a single future payment (*P* given *F*) and the stream of future payments (*P* given *A*) are tabulated in the *NCEES Handbook*, in the Engineering Economics section. Select the interest rate table for $i = 8\%$.

step 1: For equipment 1, its useful life is 10 years. Therefore, read two factor values for $n = 10$ in the table. The factor value P/F is used to convert the future salvage value to the present income value. The factor value P/A is used to convert the annual maintenance values to a single present expense.

Calculate the present worth, P_1, for the first equipment option.

$$P_1 = \begin{pmatrix} \text{initial cost} \\ \quad + \text{annual maintenance}(P/A, 8\%, 10) \\ \quad - \text{salvage value}(P/F, 8\%, 10) \end{pmatrix}$$

$$= \$50{,}000 + (\$15{,}000)(6.7101) - (\$5000)(0.4632)$$

$$= \$148{,}335.50$$

step 2: The present worth, P_2, for the second equipment option is

$$P_2 = \begin{pmatrix} \text{initial cost} \\ \quad + \text{annual maintenance}(P/A, 8\%, 15) \\ \quad - \text{salvage value}(P/F, 8\%, 15) \end{pmatrix}$$

$$= \$75{,}000 + (\$10{,}000)(8.5595) - (\$12{,}000)(0.3152)$$

$$= \$156{,}812.60$$

step 3: Select the equipment that costs less. Equipment 1 costs less than equipment 2. Therefore, it is a better choice. Determine the difference in cost.

$$\begin{aligned} \text{savings} &= P_2 - P_1 \\ &= \$156{,}812.60 - \$148{,}335.50 \\ &= \$8477.10 \quad (\$8500) \end{aligned}$$

The answer is (A).

22. This is an optimization problem in which the effect of each toll fee increment must be evaluated. From the Differential Calculus section of Mathematics in the *NCEES Handbook*, the function $y = f(x)$ is a maximum for $x = a$ if $f'(a) = 0$ and $f''(a) < 0$.

Let x be the number of \$0.25 fee increases needed to fully optimize the fee.

$$\begin{aligned} \text{new fee} &= 3.00 + 0.25x \\ \text{no. of motorists} &= 20{,}000 - 1000x \\ \text{new income, } I &= (3.00 + 0.25x)(20{,}000 - 1000x) \\ &= 60{,}000 - 3000x + 5000x - 250x^2 \\ &= 60{,}000 + 2000x - 250x^2 \end{aligned}$$

$$\begin{aligned} f'(I) &= \frac{dI}{dx} \\ &= 0 + 2000 - 500x = 0 \\ x &= 4 \end{aligned}$$

Check the second condition.

$$f''(I) = -500 < 0$$

Both maxima conditions are met. The income is maximized when the authority increases the charge by $4 \times \$0.25 = \$1.00$. A graph of the optimal toll increase versus income is shown.

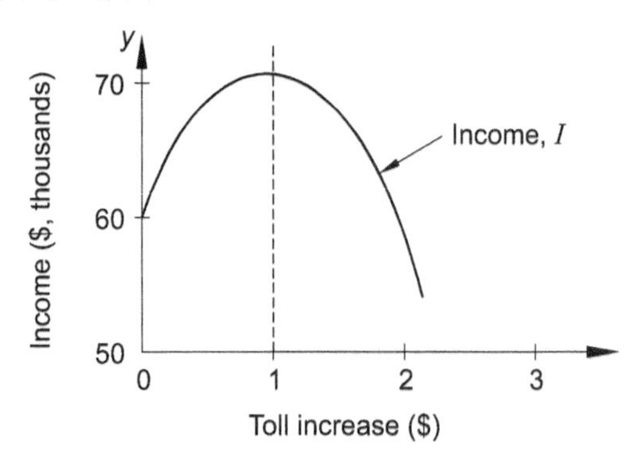

The answer is (A).

23. The life-cycle cost equations for alternatives A and B (in millions of dollars) are

$$\begin{aligned} A &= 1 + 0.1t \\ B &= 1.8 + 0.065t \end{aligned}$$

Let t be the number of years after construction. The break-even point will occur when the values of the two alternatives are equal. Set the two equations equal to each other, and solve for t.

$$\begin{aligned} 1 + 0.1t &= 1.8 + 0.065t \\ t &= \frac{1.8 - 1}{0.1 - 0.065} = 22.86 \end{aligned}$$

After 22.86 years of service life, alternative B becomes more economical, as shown.

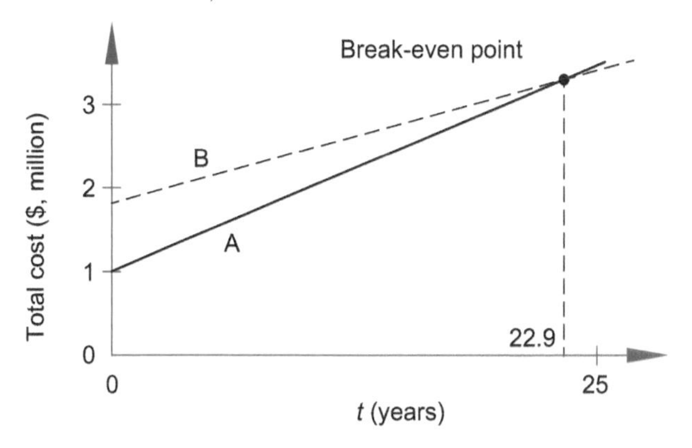

As the expected life cycle of the project is 25 years, alternative B should be used.

The answer is (B).

24. The problem involves a single lump-sum investment of \$100,000 now. The investment will lose its value at a rate of 10% compounded annually. Therefore, the problem requires converting the present payment value P to its future worth F after n interest periods at an interest rate of I per interest period. However, the interest rate is negative, and the value is declining since the equipment value is depreciating.

$$P = \$100{,}000$$
$$n = 10 \text{ yr}$$
$$i = -10\%/\text{yr}$$

There is no factor table provided in the *NCEES Handbook* for depreciation or declining rates. Therefore, use the relationship

$$\begin{aligned} F &= P(1+i)^n \\ &= \$100{,}000(1-0.1)^{10} \\ &= \$34{,}868 \quad (\$35{,}000) \end{aligned}$$

The answer is (B).

25. Draw the timeline of cash flow as shown.

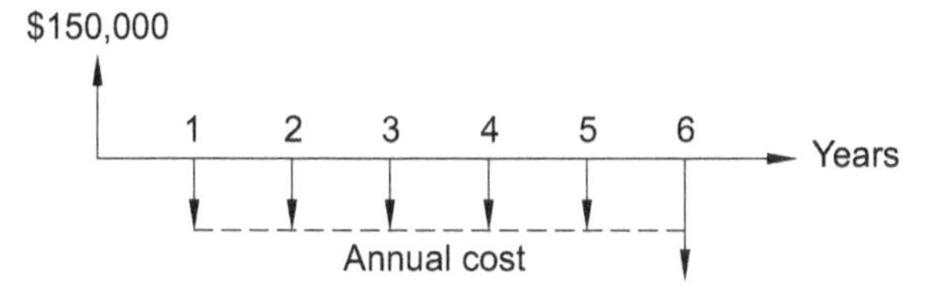

Use the factor tables given in the Engineering Economics section of the *NCEES Handbook* to determine the annual cost.

$$\begin{aligned} \text{annual cost} &= \begin{pmatrix} \$150{,}000(A/P, 8\%, 6) \\ -\$10{,}000(A/F, 8\%, 6) \end{pmatrix} \\ &= (\$150{,}000)(0.2163) - (\$10{,}000)(0.1363) \\ &= \$31{,}082 \quad (\$31{,}000) \end{aligned}$$

The answer is (C).

26. The circuit has two loops. Assume loop currents as I_1 and I_2, as shown. More than one path exists for current to flow in the circuit.

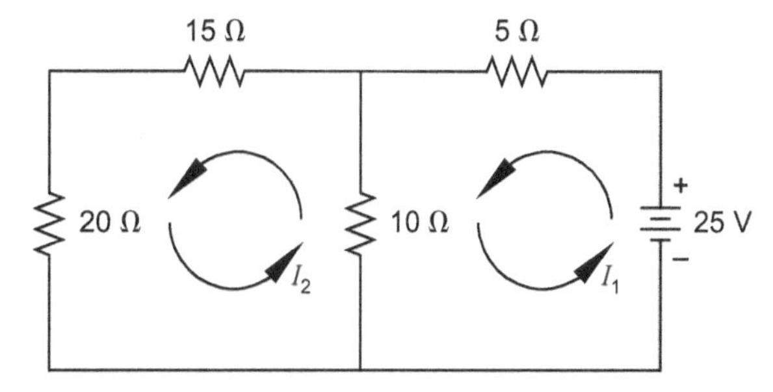

Considering the conservation of energy in electrical circuits, use Kirchhoff's voltage law (KVL). KVL states that the sum of the rises and drops in voltage around any closed path in an electrical circuit must be zero. In other words,

$$\sum V_{\text{rises}} = \sum V_{\text{drops}}$$

The loops have resistors which absorb energy. Apply Ohm's law.

$$V = I \times R$$

V is the voltage in volts, I is the current in amperes, and R is the resistance in ohms. Write an equation for voltage equilibrium for each loop in voltage.

For loop 1,

$$\begin{aligned} \sum V_1 &= 0 \\ &= -25 + 5I_1 + 10I_1 - 10I_2 = 0 \\ 15I_1 - 10I_2 &= 25 \end{aligned}$$

For loop 2,

$$\begin{aligned} \sum V_2 &= -10I_1 + 10I_2 + 15I_2 + 20I_2 = 0 \\ 10I_1 + 45I_2 &= 0 \\ I_1 &= 4.5I_2 \end{aligned}$$

Solving the simultaneous equations gives $I_1 = 1.96$ A and $I_2 = 0.43$ A.

Both current values are positive, which indicates that the assumed current directions are correct. The current in 10 Ω resistor is the difference between the two currents.

$$\begin{aligned} I_{10\ \Omega} &= I_1 - I_2 \\ &= 1.96\ \text{A} - 0.43\ \text{A} \\ &= 1.53\ \text{A} \end{aligned}$$

The answer is (B).

27. A complex circuit that has many electrical components can generally be modeled by a single source and single resistor in series. The simplified circuit is called the Thevenin equivalent circuit, as shown.

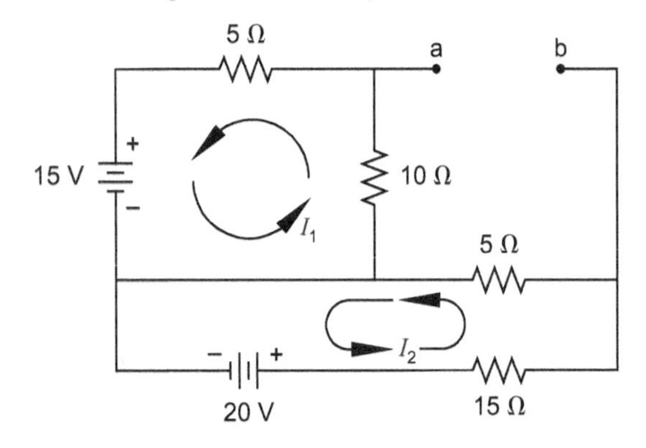

Let V_{eq} be the equivalent voltage across terminals a and b with 20 Ω removed from the circuit. The circuit has two loops with currents I_1 and I_2 as shown.

See the *NCEES Handbook*, Electrical and Computer Engineering section. For resistors connected in series, the current in all resistors is the same, and the equivalent resistance for n resistors in series is

$$\sum R = R_1 + R_2 + \cdots + R_n$$

Apply Ohm's law. The current is $I = V/R$. Calculate the current in each loop.

$$I_1 = \frac{15\ \text{V}}{5\ \Omega + 10\ \Omega} = \frac{15\ \text{V}}{15\ \Omega} = 1\ \text{A}$$

$$I_2 = \frac{20\ \text{V}}{5\ \Omega + 15\ \Omega} = \frac{20\ \text{V}}{20\ \Omega} = 1\ \text{A}$$

Using the conservation of energy in each loop,

$$\begin{aligned} \sum V_{\text{drop}} &= 0 \\ V_{ab} + (5\ \Omega)I_2 - (10\ \Omega)I_1 &= 0 \\ V_{ab} + (5\ \Omega)(1\ \text{A}) - (10\ \Omega)(1\ \text{A}) &= 0 \\ V_{ab} + 5\ \text{V} - 10\ \text{V} &= 0 \\ V_{ab} - 5\ \text{V} &= 0 \\ V_{ab} &= 5\ \text{V} \end{aligned}$$

The open circuit voltage V_{eq} equals V_{ab}.

The answer is (A).

28. See the *NCEES Handbook*, Electrical and Computer Engineering section. KCL applies to the current at any closed surface such as a node or junction. KCL does not apply to currents in a circuit. Therefore, statement B is correct, and statement A is not.

KCL states that the sum of all incoming currents equal the sum of all outgoing currents at a node. In other words, the algebraic sum of currents at a node is zero. Therefore, statement D is correct, and statement C is incorrect.

In applying KCL, it is necessary to assume a direction of current as positive. It is for a reference. The assumed direction may be clockwise or counterclockwise. Therefore, statement E is incorrect.

The answer is (B) and (D).

29. See the *NCEES Handbook*, Electrical and Computer Engineering section. According to the first law of electrostatics, like charges repel each other and opposite charges attract each other. According to the second law, known as Coulomb's law, the electrostatic force on charge 2 due to charge 1 is

$$F_2 = \frac{Q_1 Q_2}{4\pi\varepsilon r^2} a_{r12}$$

In this equation, Q_i is the ith point charge, r is the distance between charges 1 and 2, a_{r12} is a unit vector directed from 1 to 2, and ε is the permittivity of the medium.

Since all charges in this problem are in the same medium and in a line, the charge equations between A and B and between C and B are simplified to determine x, the distance AB.

$$F_{AB} = \frac{Q_A Q_B}{4\pi x^2}$$

$$F_{CB} = \frac{Q_C Q_B}{4\pi(1.58 - x)^2}$$

Set F_{AB} and F_{CB} as equal and solve for x.

$$\frac{Q_{\text{A}}Q_{\text{B}}}{4\pi x^2} = \frac{Q_{\text{C}}Q_{\text{B}}}{4\pi(1.58-x)^2}$$
$$Q_{\text{A}}Q_{\text{B}}4\pi(1.58-x)^2 = Q_{\text{C}}Q_{\text{B}}4\pi x^2$$
$$\frac{(1.58-x)^2}{x^2} = \frac{Q_C}{Q_A} = \frac{2\text{ C}}{6\text{ C}}$$
$$2x^2 - 9.48x + 7.489 = 0$$
$$x = 1.002\text{ m} \quad (100\text{ cm})$$

The answer is (C).

30. See the *NCEES Handbook*, Electrical and Computer Engineering section. The AC motor's synchronous speed, n_s, is expressed as

$$n_s = \frac{120f}{p}$$

In this equation, f is the line voltage frequency in hertz, and p is the number of poles.

Apply the formula in the following two steps.

step 1: Determine the frequency of the three-phase current produced by the alternator.

$$f = \frac{pn_s}{120} = \frac{(8)(750\text{ rpm})}{120} = 50\text{ Hz}$$

step 2: Based on the frequency, determine the speed of the three-phase induction motor.

$$n_s = \frac{120f}{p} = \frac{(120)(50\text{ Hz})}{6} = 1000\text{ rpm}$$

step 3: The difference between synchronous speed and actual speed is called *slip*. The slip for an induction motor is

$$s = \frac{n_s - n}{n_s}$$

Rewrite the above equation to determine the actual rotational speed, n.

$$n = n_s(1-s) = 1000(1-0.04) = 960\text{ rpm}$$

The answer is (D).

31. A force is a vector possessing magnitude and direction. The forces are added according to the parallelogram law. A parallelogram with sides A and B represents the resultant force vector, as shown.

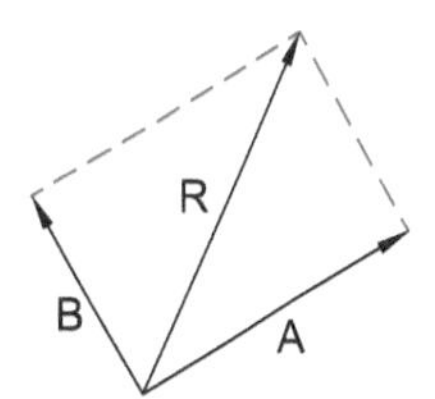

Option D correctly represents the directions and magnitudes of the force vectors and their resultant.

The answer is (D).

32. See the *NCEES Handbook*, Statics: Moments (Couples) section. A couple is defined as a system of two equal and opposite forces that are parallel to each other. As such, the forces are nonconcurrent. Therefore, option C is correct, and options A, B, and D are incorrect.

The answer is (C).

33. See the Statics section of the *NCEES Handbook*. Two conditions must be met for equilibrium.

$$\sum F_n = 0$$
$$\sum M_n = 0$$

The block weighs 1200 lbf. The weight acts vertically down at the centroid of the block, which is located midway between points A and B. The weight is equally split between the two corners A and B. Corner A bears against the surface below, and corner B is carried by the crowbar.

Calculate the load on the tip of the crowbar at B as

$$F_{\text{B}} = 0.5(1200\text{ lbf}) = 600\text{ lbf}$$

The crowbar has two arms, BC and CD. Assuming counterclockwise moment is positive, consider the moment at point C.

$$\sum M_C = 0$$
$$F_B(\text{BC}) - F_D(\text{CD}) = 0$$
$$600 \text{ lbf}(2 \text{ ft}) - F_D(6 \text{ ft}) = 0$$
$$F_D = \frac{600 \text{ lbf}(2 \text{ ft})}{6 \text{ ft}} = 200 \text{ lbf}$$

The answer is (B).

34. The tower is a plane determinate truss. See the Statics section of the *NCEES Handbook*. Two equilibrium conditions must be met.

$$\sum F_x = 0$$
$$\sum F_y = 0$$

For the pole AD to remain plumb, the horizontal components of the tension forces in each cable must be equal. Using the free-body diagram at joint A, resolve the forces in the x- and y-directions.

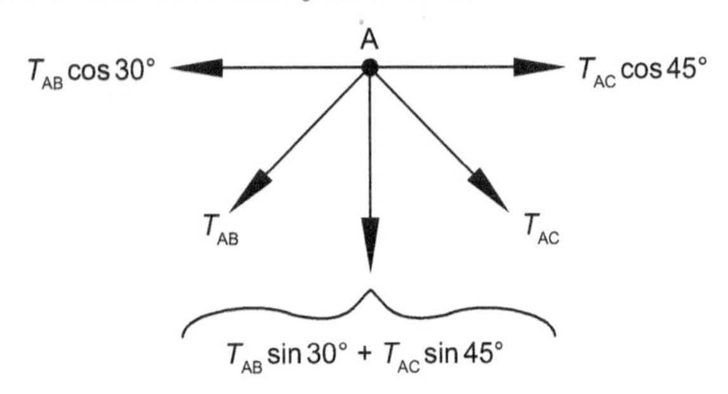

$$\sum F_x = 0$$
$$T_{AB}\cos 30° - T_{AC}\cos 45° = 0$$
$$T_{AB}\cos 30° = T_{AC}\cos 45°$$
$$100(0.866) = (T_{AC})(0.707)$$
$$T_{AC} = 122.5 \text{ N} \quad (120 \text{ N})$$

The answer is (B).

35. See the *NCEES Handbook*, Civil Engineering: Stability, Determinacy, and Classification of Structures. A truss is considered stable and indeterminate if

$$m + r > 2j$$

m is the number of members, j is the number of joints, and r is the number of reactions. From the above inequality, the degree of indeterminacy for a truss structure can be determined as

$$\text{Degree of indeterminacy} = m + r - 2j$$

The tower is a planar (2-D) truss with two supports. Each support can develop horizontal and vertical reactions.

$$r = 2(2) = 4$$
$$m = 17$$
$$j = 9$$

The degree of indeterminacy of the tower is

$$m + r - 2j = 17 + 4 - 2(9) = 3$$

The answer is (C).

36. The center of gravity (CG) of a body, also called its centroid, is defined as a point where the total area of the body acts. See the *NCEES Handbook*, Statics: Centroids of Masses, Areas, Lengths, and Volumes section, for the formula to compute the CG of an area. However, the solution does not require any calculations to determine the CG of the area.

step 1: Draw a sketch of the disk with the hole, as shown with the x- and y-axes. To locate the CG of an area, its x- and y-coordinates are needed. Due to symmetry about x-axis, the y-coordinate of the CG is zero, and thus the CG lies on the x-axis.

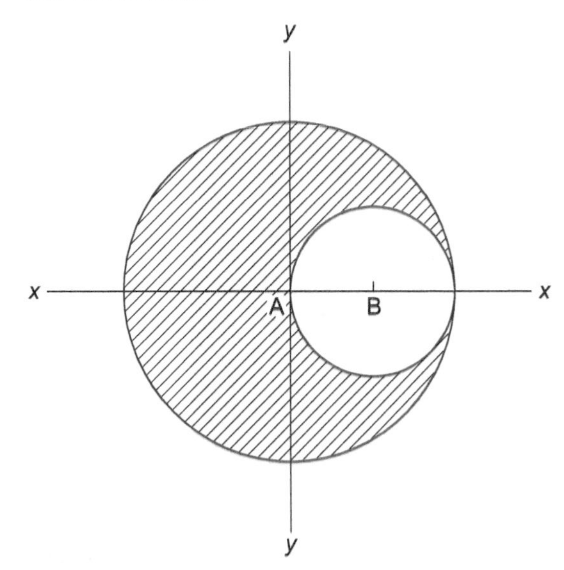

step 2: It is known that the CG of a circle lies at its center. Therefore, the CG of a 12 in radius solid disk lies at point A, and the CG of a 6 in radius void lies at point B.

step 3: After the 6 in radius circular is cut out, there is more area left of point A than on the right side of point A. Therefore, the CG of the composite area shifts to the left of point A, along the x-axis.

The answer is (C).

37. The I-shaped beam can be considered in terms of three rectangular shapes.

I-shaped section = Rectangle A – Rectangle B – Rectangle C

I-shaped section = Rectangle A – Rectangle (B + C)

This is shown as

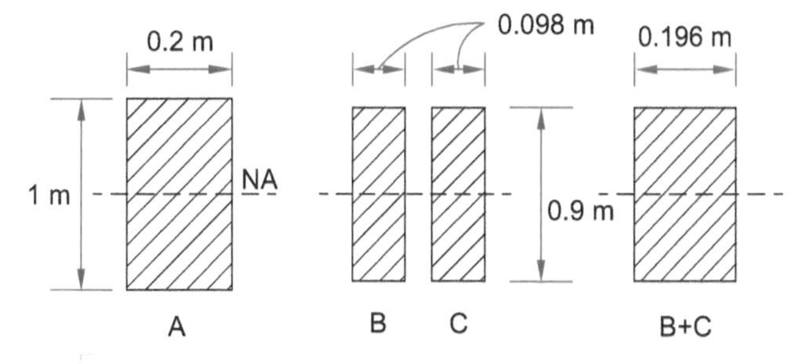

The moment of inertia of a rectangular section about its axis is determined using the formula

$$I = \frac{bd^3}{12}$$

Substituting the sizes of the two rectangles into the equation,

$$\begin{aligned} I &= \frac{(0.2\ \text{m})(1\ \text{m})^3}{12} - \frac{(0.196\ \text{m})(0.9\ \text{m})^3}{12} \\ &= 4.76 \times 10^{-3}\ \text{m}^4 \end{aligned}$$

Alternative method: Use the moment of inertia parallel-axis theorem.

The answer is (D).

38. The belt friction formula is given in the *NCEES Handbook*, Statics: Belt Friction section.

The force F_1 in the formula represents the tensile force on the pulling side. It is always larger than F_2. Therefore, statement A is correct, and statement B is incorrect.

The formula applies to friction problems where the angle of contact may exceed 360° (2π rad). If the rope is wrapped n times around a post, then angle θ becomes $2\pi n$. Statement C is true.

The formula applies to friction problems involving band brakes. In band brakes, the band remains fixed, and the drum has impending rotation. Statement D is false.

The formula is applicable to problems involving ropes wrapped around a capstan or post. In this case, the capstan remains fixed, and the angle θ may exceed 2π rad. Statement E is false.

The answer is (A) and (C).

39. The position of the applied load W is not given. The load can be applied anywhere between points A and B to determine the controlling condition. For the ladder to be in equilibrium, three conditions must be met, as given in the *NCEES Handbook*.

$$\sum F_x = 0$$
$$\sum F_y = 0$$
$$\sum M_B = 0$$

For the general condition, when both the wall and floor can develop friction and the load is located at a distance x, the reactions are as shown.

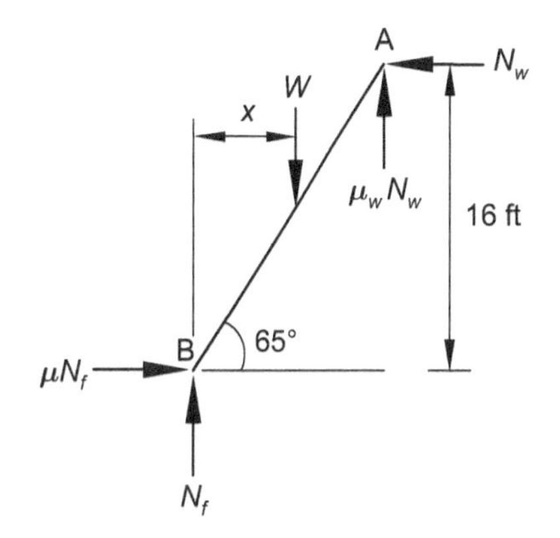

step 1: Apply the three conditions to the forces shown.

$$\sum F_x = N_w - \mu_f N_f = 0$$
$$\sum F_y = N_f - \mu_w N_w - 200\ \text{lbf} = 0$$
$$\begin{aligned} \sum M_B &= N_w(16\ \text{ft}) - \mu_w N_w \cot 65° - (200\ \text{lbf})(x\ \text{ft}) \\ &= 0 \end{aligned}$$

step 2: It is known that $\mu_w = 0$. Simplify the above equilibrium conditions.

$$\begin{aligned} N_w &= \mu_f N_f \\ N_f &= 200\ \text{lbf} \\ N_w(16\ \text{ft}) &= (200\ \text{lbf})x \end{aligned}$$

step 3: Solve for μ_f.

$$\begin{aligned} N_w &= \mu_f(200\ \text{lbf}) \\ \mu_f(200\ \text{lbf})(16\ \text{ft}) &= (200\ \text{lbf})x \\ \mu_f &= \frac{x}{16\ \text{ft}} \end{aligned}$$

step 4: The above relationship shows that the required friction factor depends on the horizonal distance of the applied weight. It is zero when the applied load is at point B, and it is at its maximum when the load is at point A. At its maximum, the friction factor is

$$\begin{aligned} \mu_f &= \frac{x_{\text{A}}}{16 \text{ ft}} \\ &= \cot 65° \\ &= 0.47 \quad (0.5) \end{aligned}$$

Thus, the minimum coefficient of friction needed at the floor to keep the ladder stable is 0.5.

The answer is (D).

40. Resolve the applied load F at point C into its x- and y-components as shown.

$$\begin{aligned} F_x &= F\cos 30° \\ F_y &= F\sin 30° \end{aligned}$$

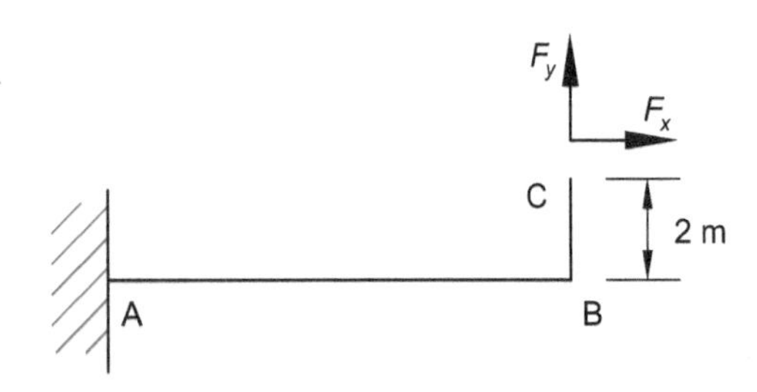

Move the force system from C to B. Use the principle that any force acting at point C can be moved to point B provided that a couple is added with a moment equal to the moment of F about B.

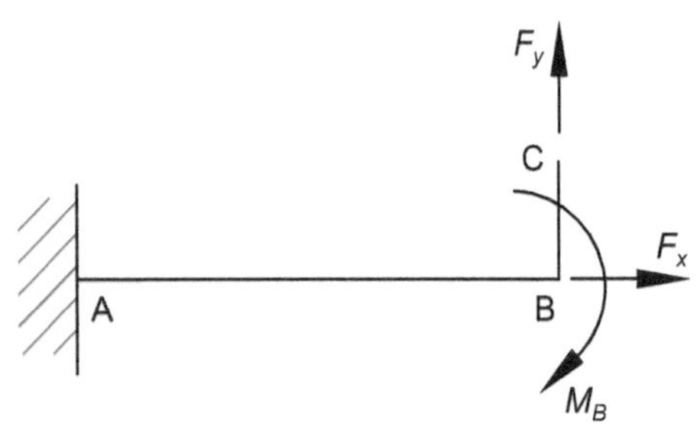

Let counterclockwise moments be positive. The associated moment is

$$\begin{aligned} M_B &= F_x(-2) + F_y(0) \\ &= -2F_x \end{aligned}$$

Similarly, move the force system from B to A.

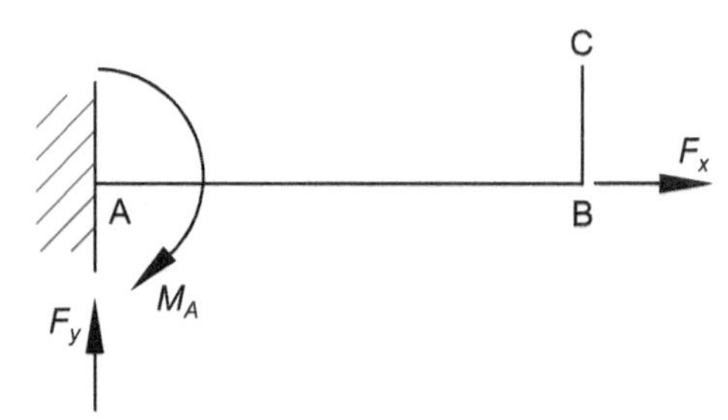

The resulting moment at support A is calculated as

$$\begin{aligned} M_A &= F_y(9 \text{ m}) + M_B \\ &= 9F_y - 2F_x \\ &= 9F\sin 30° - 2F\cos 30° \\ &= F(9\sin 30° - 2\cos 30°) \\ &= (2.77 \text{ m})F \\ &= (2.77 \text{ m})(10 \text{ kN}) \\ &= 27.7 \text{ kN·m} \quad (28 \text{ kN·m, ccw}) \end{aligned}$$

The answer is (B).

41. The stone's initial velocity, u, is zero. The following information is given.

$$\begin{aligned} u &= 0 \text{ m/s} \\ a &= g = 9.81 \text{ m/s}^2 \\ t &= 2.9 \text{ s} \end{aligned}$$

Determine the distance traveled, s, as

$$\begin{aligned} s &= ut + \tfrac{1}{2}at^2 \\ &= \left(0 \ \frac{\text{m}}{\text{s}}\right)(2.9 \text{ s}) + \frac{1}{2}\left(9.81 \ \frac{\text{m}}{\text{s}^2}\right)(2.9 \text{ s})^2 \\ &= 41.25 \text{ m} \quad (41 \text{ m}) \end{aligned}$$

The answer is (C).

42. Given:

$$\begin{aligned} \mu &= 0.7 \\ N &= 145 \text{ lbf} \end{aligned}$$

Determine the impending sliding force as

$$\begin{aligned} F &= \mu N \\ &= 0.7 \ (145 \text{ lbf}) \\ &= 101.5 \text{ lbf} \quad (102 \text{ lbf}) \end{aligned}$$

The answer is (B).

43. The change in kinetic energy is the work done in accelerating the flywheel from ω_1 to ω_2. The work-energy formula is

Torque $\times$ Angle turned (rad) = Increase in kinetic energy

$$T \times \theta = \frac{1}{2}\left(I\omega_2^2 - I\omega_1^2\right)$$

Given:

$$T = 0.1 \text{ kg·m}^2/\text{s}^2$$

$$\begin{aligned} \theta &= (10 \text{ turns})\left(2\pi \ \frac{\text{rad}}{\text{turn}}\right) \\ &= 20\pi \text{ rad} \end{aligned}$$

Therefore, $\omega_1 = 0$.

$$\begin{aligned} T\theta &= \frac{I\omega_2^2}{2} \\ \omega_2 &= \sqrt{\frac{2T\theta}{I}} \\ &= \sqrt{\frac{2\left(0.1 \text{ kg·m}^2/\text{s}^2\right)(20\pi \text{ rad})}{3.14 \text{ kg·m}^2}} \\ &= 2 \text{ rad/s} \end{aligned}$$

The answer is (B).

44. This is a plastic case wherein the vehicles stick together after the collision. Use the impact equation for conservation of momentum given in the Dynamics section of the *NCEES Handbook.*

$$\begin{aligned} m_1 v_1 + m_2 v_2 &= m_1 v'_1 + m_2 v'_2 \\ &= v_{\text{comb}}(m_1 + m_2) \end{aligned}$$

In this case, the velocities are equal but in opposite directions. Therefore, $v_2 = -v_1$. Assume the eastbound velocity (v_1) is positive.

$$\begin{aligned} v_1(m_1 - m_2) &= v_{\text{comb}}(m_1 + m_2) \\ v_{\text{comb}} &= v_1 \frac{m_1 - m_2}{m_1 + m_2} \\ &= v_1 \frac{g(w_1 - w_2)}{g(w_1 + w_2)} \\ &= v_1 \frac{6000 \text{ lbf} - 15{,}000 \text{ lbf}}{6000 \text{ lbf} + 15{,}000 \text{ lbf}} \\ &= \frac{\left(30 \ \frac{\text{m}}{\text{h}}\right)(-9000 \text{ lbf})}{21{,}000 \text{ lbf}} = -12.9 \text{ mph} \\ &= -12.9 \text{ mph} \quad (13 \text{ mph to the west}) \end{aligned}$$

The answer is (C).

45. Use units of ft-lbf. The spring undamped frequency, ω_n, can be determined in terms of the static deflection of the system, δ_{st}.

$$\omega_n = \sqrt{\frac{k}{m}}$$

The stiffness, k, is the force required to stretch a spring per unit length.

Use the static spring-displacement formula for constant force given in the Dynamics section of the *NCEES Handbook.*

$$mg = k\delta_{\text{st}}$$

Since $mg = W$, the formula reduces to

$$\begin{aligned} k &= \frac{W}{\delta_{\text{st}}} \\ &= \frac{4 \text{ lbf}}{6 \text{ in}}\left(\frac{12 \text{ in}}{1 \text{ ft}}\right) \\ &= 8 \text{ lbf/ft} \end{aligned}$$

$$g = 32.2 \ \frac{\text{ft}}{\text{sec}^2}$$

$$m = \frac{W}{g} = \frac{4 \text{ lbf}}{32.2 \ \frac{\text{ft}}{\text{sec}^2}} \approx \frac{1}{8} \text{ slugs}$$

$$\omega_n = \sqrt{\frac{8\ \frac{\text{lbf}}{\text{ft}}}{\frac{1}{8}\ \text{slugs}}} = 8\ \text{rad/sec}$$

The answer is (D).

46. The average force equals the change in momentum per unit time. It is expressed as

$$F = \frac{d(m\text{v})}{dt} = ma$$

In this case, the ball decelerated to a complete stop ($\text{v}_2 = 0$).

$$a = \frac{\text{v}_2 - \text{v}_1}{t} = \frac{30\ \frac{\text{m}}{\text{s}} - 0\ \frac{\text{m}}{\text{s}}}{0.1\ \text{s}} = 300\ \text{m/s}^2$$

Therefore,

$$F = \frac{d(m\text{v})}{dt} = ma = (0.149\ \text{kg})\left(300\ \frac{\text{m}}{\text{s}^2}\right) = 44.7\ \text{N}$$

The answer is (B).

47. The total kinetic energy (KE) of a roller includes its KE of translation and KE of rotation.

$$E_{\text{total}} = \tfrac{1}{2}m\text{v}^2 + \tfrac{1}{2}I\omega^2$$

The kinetic energy of translation (motion) depends on the mass of the entire body in motion and its velocity.

$$m = 10{,}000\ \text{kg}$$

$$\text{v} = 10\ \frac{\text{km}}{\text{h}} = \frac{10{,}000\ \frac{\text{m}}{\text{h}}}{3600\ \frac{\text{s}}{\text{h}}} = 2.78\ \text{m/s}$$

$$E_1 = \tfrac{1}{2}m\text{v}^2 = \frac{1}{2}(10{,}000\ \text{kg})\left(2.78\ \frac{\text{m}}{\text{s}}\right)^2 = 38{,}580\ \text{N·m}$$

The kinetic energy of rotation depends on the mass moments of inertia of the rotating parts and their angular velocity. In this problem, the masses of the front-axle wheel and rear-axle wheel and their angular velocities are equal.

$$I = \sum Mr^2 = (5000\ \text{kg} + 5000\ \text{kg})(0.5\ \text{m})^2 = 2500\ \text{kg·m}^2$$

The wheel radius is

$$r = \frac{1.1\ \text{m}}{2} = 0.55\ \text{m}$$

The angular velocity is

$$\omega = \frac{\text{v}}{r} = \frac{2.78\ \frac{\text{m}}{\text{s}}}{0.55\ \text{m}} = 5.05\ \text{rad/s}$$

$$E_2 = \tfrac{1}{2}I\omega^2 = \frac{1}{2}(2500\ \text{kg·m}^2)\left(5.05\ \frac{\text{rad}}{\text{s}}\right)^2 = 31{,}878\ \text{N·m}$$

$$E_{\text{total}} = E_1 + E_2 = 38{,}580\ \text{N·m} + 31{,}878\ \text{N·m} = 70{,}465\ \text{N·m}\quad (70{,}000\ \text{N·m})$$

The answer is (D).

48. The deceleration, a, and coefficient of friction, μ, are related as

$$a = \mu g$$

Determine the coefficient of friction, μ, between the road and the vehicle as

$$\mu = \frac{a}{g} = \frac{13 \ \frac{\text{ft}}{\text{sec}^2}}{32.2 \ \frac{\text{ft}}{\text{sec}^2}} = 0.4$$

The answer is (B).

49. In order for the block to move up the slope, the applied force must overcome two resisting forces: friction and gravity. Since the applied force is zero when time is zero, the block will start moving sometime later. Let time t_1 be the moment that the block starts to move. The block's velocity will increase with time, as shown.

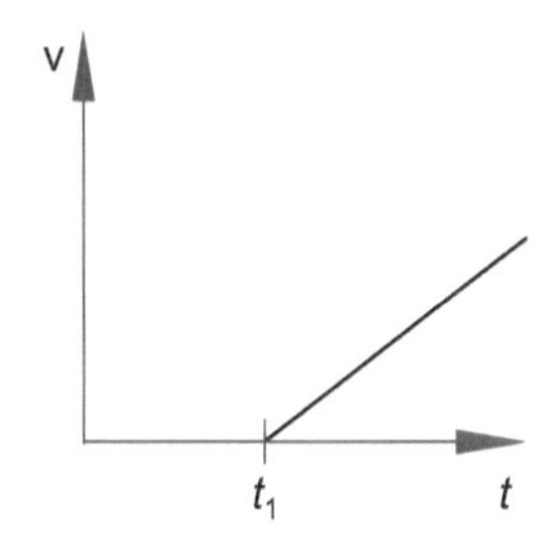

step 1: Use the *NCEES Handbook*, Dynamics section. Draw a free-body diagram of the forces acting on the block, resolving the forces in the x- and y-directions.

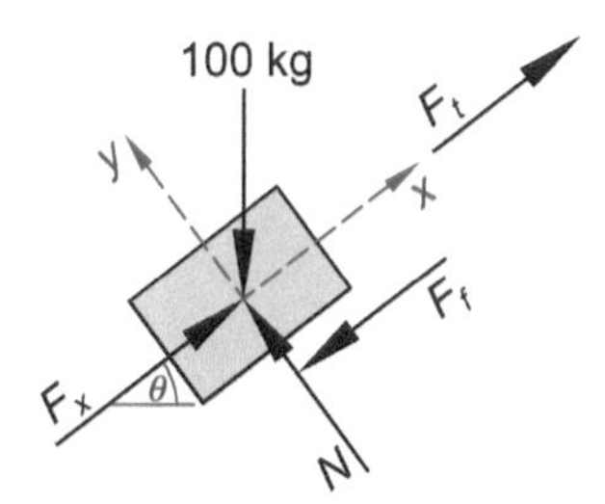

Calculate the tangential force, F_x, and the normal force, N, as

$$F_x = (100 \text{ kg})\left(\frac{9.81 \text{ N}}{1 \text{ kg}}\right)\sin\theta = (100 \text{ kg})\left(\frac{9.81 \text{ N}}{1 \text{ kg}}\right)\left(\frac{3}{5}\right) = 589 \text{ N}$$

$$N = (100 \text{ kgf})\left(\frac{9.81 \text{ N}}{1 \text{ kgf}}\right)\cos\theta = (100 \text{ kgf})\left(\frac{9.81 \text{ N}}{1 \text{ kgf}}\right)\left(\frac{4}{5}\right) = 785 \text{ N}$$

Calculate the friction force between the block and the sliding surface as

$$F_f = \mu N = (0.2)(785 \text{ N}) = 157 \text{ N}$$

The force-time diagram shows that the applied force increases linearly with time. The impending motion of the block will initiate when the applied force equals the sum of resisting forces.

$$F_t = \left(400 \ \frac{\text{N}}{\text{s}}\right)t$$

$$t = \frac{F_t}{400 \ \frac{\text{N}}{\text{s}}} = \frac{157 \text{ N} + 589 \text{ N}}{400 \ \frac{\text{N}}{\text{s}}} = 1.87 \text{ sec}$$

The answer is (C).

50. See the instantaneous center (IC) method given in the Dynamics section of the *NCEES Handbook*. The rod AB rotates about point A, and the rod CD rotates about point D. The rod BC rotates about its IC, located at C as shown.

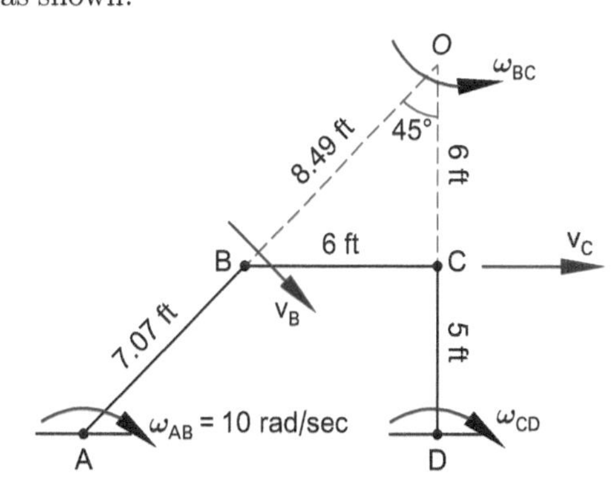

Determine the linear velocity at point B, v_B, as

$$\begin{aligned} v_B &= r_{AB}\omega_{AB} \\ &= (7.07\text{ ft})\left(10\ \frac{\text{rad}}{\text{sec}}\right) \\ &= 70.7\text{ ft/sec} \quad (71\text{ ft/sec}) \end{aligned}$$

The answer is (D).

51. The material properties of steel are given in the Typical Material Properties table in the Mechanics of Materials: Material Properties section of the *NCEES Handbook*. From the Units and Conversion Factors table of the *NCEES Handbook*,

$$\begin{aligned} 1\text{ Pa} &= 1\ \frac{\text{N}}{\text{m}^2} \\ 1\text{ kPa} &= 1\ \frac{\text{kN}}{\text{m}^2} \end{aligned}$$

The Young's modulus of steel is

$$\begin{aligned} E &= 200\text{ GPa} \\ &= 2\times 10^8\text{ kPa} \\ &= 2\times 10^8\ \frac{\text{kN}}{\text{m}^2} \end{aligned}$$

The cross-sectional area of the bar is

$$\begin{aligned} A &= \frac{\pi D^2}{4} \\ &= \frac{\pi(0.03\text{ m})^2}{4} \\ &= 7.07\times 10^{-4}\text{ m}^2 \end{aligned}$$

The elongation of the bar, ΔL, is

$$\begin{aligned} \Delta L &= \frac{PL}{AE} \\ &= \frac{(300\text{ kN})(1\text{ m})}{(7.07\times 10^{-4}\text{ m}^2)\left(2\times 10^8\ \frac{\text{kN}}{\text{m}^2}\right)} \\ &= 2.12\text{ mm} \quad (2\text{ mm}) \end{aligned}$$

The answer is (B).

52. The bolt is in double shear. The bolt shear is distributed over two cross-sectional areas of the bolt.

$$A = \frac{\pi d^2}{4} = \frac{\pi(1\text{ in})^2}{4} = 0.785\text{ in}^2$$

Determine the shear stress, τ, in the bolt as

$$\begin{aligned} \tau &= \frac{V}{2A} \\ &= \frac{16\text{ kips}}{(2)(0.785\text{ in}^2)} \\ &= 10.19\text{ kips/in}^2 \quad (10\text{ ksi}) \end{aligned}$$

The answer is (B).

53. I is the moment of inertia of the section about its weak axis (i.e., the y-axis). Determine the moment of inertia of a rectangular section bending about its y-axis, as given in the Statics section of the *NCEES Handbook*.

$$\begin{aligned} I_y &= \frac{b^3h}{12} \\ &= \frac{(6\text{ cm})^3(6\text{ cm})}{12} \\ &= 108\text{ cm}^4 = 1.08\times 10^{-6}\text{ m}^4 \end{aligned}$$

The modulus of elasticity for steel is

$$E = 200\text{ GPa} = 200\times 10^9\text{ Pa} = 200\times 10^9\text{ N/m}^2$$

For a pinned column loaded concentrically, use the Euler equation to determine the buckling capacity, P_{cr}.

$$\begin{aligned} P_{cr} &= \frac{\pi^2 EI}{L^2} \\ &= \frac{\pi^2\left(200\times 10^9\ \frac{\text{N}}{\text{m}^2}\right)(1.08\times 10^{-6}\text{ m}^4)}{(2\text{ m})^2} \\ &= 533\times 10^3\text{ N} \\ &= 533\text{ kN} \quad (530\text{ kN}) \end{aligned}$$

The answer is (C).

54. The following information is given.

$$\begin{aligned} \Delta T &= 98°\text{F} - 68°\text{F} \\ &= 30°\text{F} \end{aligned}$$

$$e_{th} = 6.5\times 10^{-6}\ \frac{\text{in}}{\text{in-°F}}$$

Determine the error induced as

$$\begin{aligned} \Delta L &= e_{th}L\Delta T \\ &= \left(6.5\times 10^{-6}\ \frac{\text{in/in}}{°\text{F}}\right)(3001.20\text{ ft})(30°\text{F}) \\ &= 0.59\text{ ft} \end{aligned}$$

Due to the high temperature, the tape expanded 0.59 ft. Therefore, the measured length is less than the true length. The true distance is calculated as

$$\begin{aligned} L_{\text{true}} &= 3001.20 \text{ ft} + 0.59 \text{ ft} \\ &= 3001.79 \text{ ft} \end{aligned}$$

The answer is (C).

55. The external forces consist of the vertical reactions at beam supports. From statics, the UDL and the concentrated load, P, are equal in magnitude, and their centers of gravity coincide. As such, both loadings have their CGs at the midspan of the beam. Therefore, the beam reactions are equal for both loadings. Option A is correct.

The shear force and bending moments in the beam across the span are the internal forces. The internal forces are unique to the type of loading. See the *NCEES Handbook*, Mechanics of Materials: Simply Supported Beam Slopes and Deflections. This table shows that the maximum bending moment induced by a concentrated load P is twice the bending moment induced by the same total load distributed over the beam span. Similarly, the deflection profiles under the loading differ substantially. Therefore, options B and C are incorrect.

The answer is (A).

56. For the shear force and bending moment sign convention, use the convention given in the *NCEES Handbook*, Mechanics of Materials: Beams.

The beam loading is symmetrical and, therefore, the reactions at supports A and C are equal. The shear force at any section is defined as the sum of the reaction and applied forces on the left of the section. To determine the shear force at any point along the beam span, consider the left support A as a reference point, and select the section at any distance between A and B.

1. The shear force at the left support A equals the support reaction.
2. The shear force at the section between A and B is the sum of the support reaction at A and the applied loads. As the applied uniformly distributed load is in the direction opposite to the reaction, the shear force at all sections between A and B would decrease linearly. Therefore, options A, B, and C are incorrect.
3. The concentrated load at point B acts downward. The shear force is defined as the sum of the support reaction at A and the applied loads, so the shear force at midspan has a sharp drop, as shown in the option D.
4. By symmetry, the shear force diagram for the right side of the span, CB, is the mirror image of the shear force in the left side of the span, AB, except for the sign.

As given in the *NCEES Handbook*, the shear force is considered positive if the right portion of the beam tends to shear downward with respect to the left. In this case, the right portion of the beam tends to shear downward with respect to the left. Therefore, the shear force diagram option D has the proper sign convention.

The answer is (D).

57. The beam has two equal spans and is subjected to a UDL. The beam slope at the central support A is zero due to the symmetry of geometry and loading. Because the middle support A does not rotate, it can be considered as fixed. Therefore, the two-span beam can be visualized as two single-span propped cantilevers, as shown.

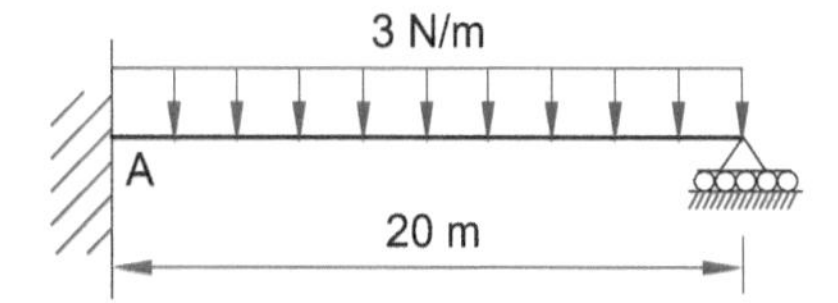

The moment at the fixed end A of a propped cantilever, as given in the problem statement, is

$$\begin{aligned} M_{\text{max}} &= \frac{wL^2}{8} \\ &= \frac{\left(3 \ \frac{\text{N}}{\text{m}}\right)(20 \text{ m})^2}{8} \\ &= 150 \text{ N·m} \end{aligned}$$

The answer is (C).

58. For steel, $E = 29{,}000$ ksi. For a 36 ksi yield stress, the yield strain, ε_y, is calculated as

$$\begin{aligned} \varepsilon_y &= \frac{\sigma_y}{E} \\ &= \frac{36 \ \frac{\text{kips}}{\text{in}^2}}{29{,}000 \ \frac{\text{kips}}{\text{in}^2}} \\ &= 0.001241 \end{aligned}$$

Substitute the value of yield strain, ε_y, to determine the elongation at yield, ΔL_y, as

$$\begin{aligned} \Delta L_y &= (0.00124)(3 \text{ ft})\left(12 \ \frac{\text{in}}{\text{ft}}\right) \\ &= 0.0447 \text{ in} \end{aligned}$$

Since specimen elongation is greater than yield elongation, the specimen went into its plastic range, causing permanent elongation. Determine the permanent elongation of the bar in inches as

$$\begin{aligned}\Delta L_{\text{permanent}} &= \Delta L_{\text{total}} - \Delta L_y \\ &= 0.20 \text{ in} - 0.0447 \text{ in} \\ &= 0.1553 \text{ in} \quad (0.155 \text{ in})\end{aligned}$$

The answer is (C).

59. The shaft is carrying a uniformly distributed load (UDL) of 250 N/m (0.25 kN/m). The bending moment under a UDL occurs at midspan as shown. The bending moment is

$$\begin{aligned}M_{\text{max}} &= \frac{wL^2}{8} \\ &= \frac{\left(250 \ \frac{\text{N}}{\text{m}}\right)(6 \text{ m})^2}{8} \\ &= 1125 \text{ N}\cdot\text{m}\end{aligned}$$

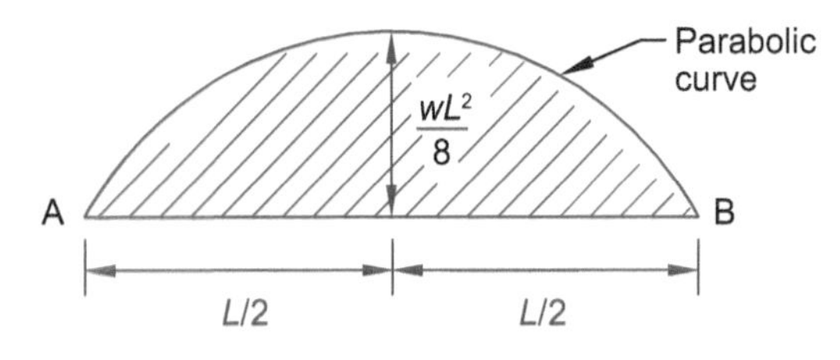

The maximum bending stress of a shaft section with moment M is

$$\sigma_{\text{max}} = \frac{Mc}{I}$$

In this equation, c is

$$\begin{aligned}c &= r \\ &= \frac{200 \text{ mm}}{2} \\ &= 100 \text{ mm} \\ &= 0.1 \text{ m}\end{aligned}$$

The moment of inertia, I, is

$$\begin{aligned}I &= \frac{\pi r^4}{4} = \frac{\pi (0.1 \text{ m})^4}{4} \\ &= 0.785 \times 10^{-4} \text{ m}^4\end{aligned}$$

Substitute I and c into the equation to determine the maximum bending stress of a shaft section σ_{max} with moment M.

$$\begin{aligned}\sigma_{\text{max}} &= \frac{Mc}{I} = \frac{(1125 \text{ N}\cdot\text{m})(0.1 \text{ m})}{0.785 \times 10^{-4} \text{ m}^4} \\ &= 1.43 \times 10^6 \ \frac{\text{N}}{\text{m}^2} \\ &= 1.43 \times 10^6 \text{ Pa} \\ &= 1.43 \text{ MPa} \quad (1.4 \text{ MPa})\end{aligned}$$

The answer is (C).

60. See the *NCEES Handbook*, Mechanics of Materials: Mohr's Circle section. The radius is best determined by using the principal stress values along the horizontal axis where the shear stress component is zero. The Mohr's circle's diameter along the x-axis equals the difference between the principal stresses. Therefore, the radius of a Mohr's circle is equal to one-half the difference of two principal stresses.

The answer is (D).

61. Use the Mohr circle method as given in the *NCEES Handbook*. Draw the circle as shown.

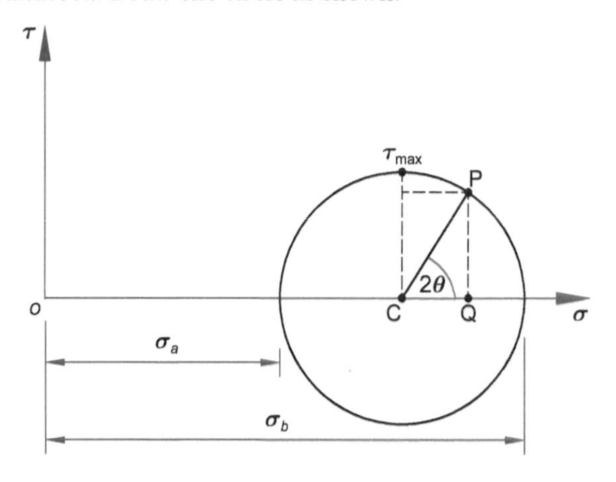

step 1: Draw the circle with the center on the normal stress. Its center is located at

$$\frac{\sigma_a + \sigma_b}{2} = \frac{100 \text{ ksi} + 50 \text{ ksi}}{2} = 75 \text{ ksi}$$

step 2: The radius of the Mohr's circle is

$$\begin{aligned}r &= \frac{100 \text{ ksi} - 50 \text{ ksi}}{2} \\ &= 25 \text{ ksi}\end{aligned}$$

The maximum shear stress equals the radius of the Mohr's circle, and it occurs on a plane where angle $\theta = 45°$ or $2\theta = 90°$.

step 3: Through C, draw a line, CP, making an angle 2θ (= 60°) with the minor (y-) axis. Draw a line, PQ, perpendicular to the x-axis. The distance PQ on the y-axis is the shear stress on the inclined plane.

$$\begin{aligned}\tau &= \text{PQ} \\ &= \text{OP}\sin 2\theta \\ &= (25 \text{ ksi})\sin 60° \\ &= 21.65 \text{ ksi} \quad (22 \text{ ksi})\end{aligned}$$

The answer is (B).

62. The standard oxidation potential for corrosion reaction for a metal has the same magnitude as its electrode potential. However, its sign is reversed. It signifies the rate at which corrosion takes place and is expressed in volts. A metal with higher oxidation potential corrodes at a higher rate. However, no method is available to measure the absolute value of the potentials. Hence, the potentials are measured under standard conditions with reference to a standard hydrogen electrode.

The answer is (C).

63. The strain energy of an element at its yield point is called the modulus of resilience of the material. It is equivalent to the work done by the load. Mathematically, it is defined as the area under the stress-strain diagram up to the yield point. For linearly elastic materials, it is the area of a right-angled triangle, as shown shaded in the stress-stress diagram.

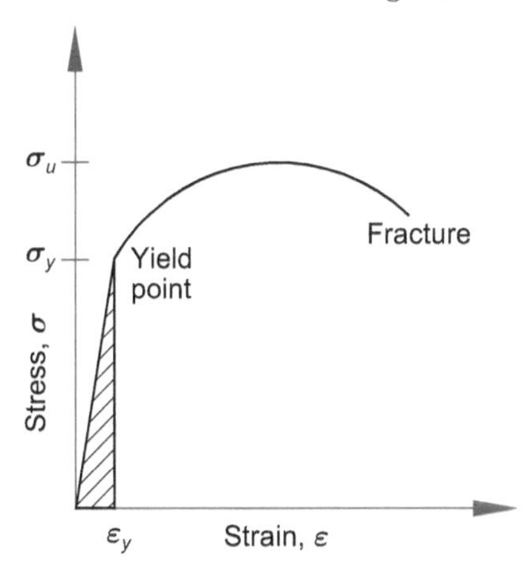

Yield stress and yield strain are given.

$$\sigma_y = 60{,}000 \text{ psi}$$
$$\varepsilon_y = 0.002 \text{ in/in}$$

Calculate the modulus of resilience as

$$\begin{aligned}U_R &= \frac{\sigma_y \varepsilon_y}{2} \\ &= \frac{(60{,}000 \text{ psi})\left(0.002 \ \frac{\text{in}}{\text{in}}\right)}{2} \\ &= 60 \text{ psi}\end{aligned}$$

The answer is (B).

64. See the *NCEES Handbook*, Mechanical Engineering: Endurance Limit for Steels. For steels with ultimate strength, S_{ut}, greater than 1400 MPa, the estimated endurance limit, S'_e, is 700 MPa.

The answer is (C).

65. In elastic range, the stress is proportional to the strain. The stress-strain relationship is linear, and its slope is called the modulus of elasticity. Therefore, option A is true. In cases in which there is no well-defined yielding point, the yield strength is defined at the 0.2% strain offset. To determine the yield strength, draw a line at 0.2% strain, running parallel to the initial stress-strain curve. The intersection of the line and the stress-strain curve is defined as the yield stress of the material, as shown.

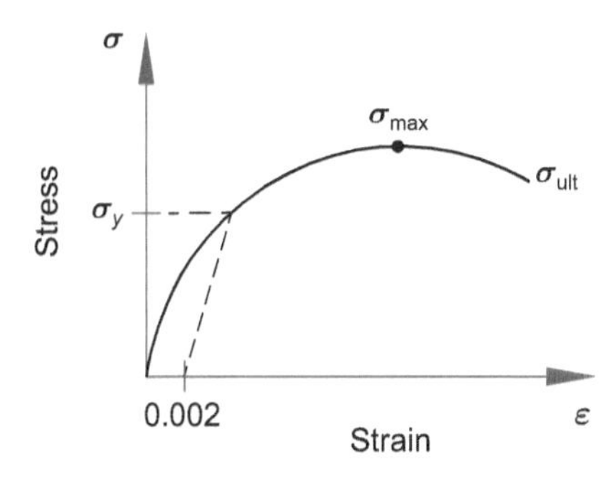

Options A, B, and C are true and are given in the Uniaxial Stress-Strain section of the *NCEES Handbook*. The true stress of a metal is

$$\text{True stress} = \frac{\text{Load}}{\text{Actual cross-sectional area}}$$

True stress differs from the engineering stress as the cross-sectional area reduces and "necks" after yielding.

The answer is (D).

66. See the *NCEES Handbook*, Material Science section. When the load is applied to an aligned fiber-reinforced composite parallel to the fibers, both matrix and fibers deform equally. This is known as isostrain condition and is expressed as

$$\begin{aligned} E_{\text{composite}} &= \sum f_i E_i \\ &= (0.7)(70.5 \text{ GPa}) + (0.3)(6.85 \text{ GPa}) \\ &= 49.35 \text{ GPa} + 2.055 \text{ GPa} \\ &= 51.405 \text{ GPa} \end{aligned}$$

Under isostrain condition, say, strain $\varepsilon = 1$, the portion of load carried by the matrix component is determined as

$$\begin{aligned} \frac{\text{Load carried by matrix}}{\text{Load carried by composite}} &= \frac{\sigma_m A_m}{\sigma_c A_c} \\ &= \frac{\sigma_m \varepsilon A_m}{\sigma_c \varepsilon A_c} \\ &= \frac{A_m}{A_c}\frac{E_m}{E_c} \\ &= \frac{f_m E_m}{E_c} \\ &= \frac{2.055 \text{ GPa}}{51.405 \text{ GPa}} \\ &= 0.0399 \quad (4.0\%) \end{aligned}$$

The answer is (A).

67. See the *NCEES Handbook*, Material Science section, Iron-Iron Carbide Phase Diagram, and use the lever rule. The eutectoid isotherm pertains to 727°C.

The abscissa in the diagram extends only to 6.7%, rather than 100%. In this material, one carbon atom bonds with three iron atoms by weight to form iron carbide. Thus, 6.7% of carbon corresponds to 100% iron carbide. At 0.77% carbon and 727°C, the phase changes.

At a temperature just higher than 727°C, austenite will have a carbon content of 0.77%, and it will transform to pearlite. The austenite consists of grains of uniform material that were formed when the steel was cooled from a liquid to a solid. For 0.28% carbon in iron carbide, use the lever rule to determine the proportions.

Primary α+ carbide: 0.02% carbon.

$$\begin{aligned} \%\text{ primary } \alpha &= \left[\frac{0.77 - 0.28}{0.77 - 0.02}\right] \times 100\% \\ &= 65.33\% \quad (65\%) \end{aligned}$$

$$\begin{aligned} \%\text{ pearlite} &= \left[\frac{0.28 - 0.02}{0.77 - 0.02}\right] \times 100\% \\ &= 34.67\% \quad (35\%) \end{aligned}$$

Calculation check: The constituents add up to 100%.

The answer is (C).

68. Copper-zinc alloys are called *brasses*. Copper-tin alloys are called *bronzes*. Copper-tin-zinc alloy is also called *gun metal*. Therefore, options A, B, and C are incorrect.

Y alloy is a nickel-containing aluminum alloy. It was a part of an experimental study to develop a new aluminum alloy to retain its strength at high temperatures in applications such as the aircraft engines. Generally, its composition is

Aluminum	92.5%
Copper	4.0%
Nickel	2.0%
Magnesium	1.5%

The answer is (D).

69. The pipe diameter, D, is 2.54 cm, or 0.0254 m. See the *NCEES Handbook*, Fluid Mechanics section, for the definition of Reynolds number. Use the kinematic viscosity and flow velocity to calculate the Reynolds number. It is a dimensionless number and is expressed as

$$\begin{aligned} \text{Re} &= \frac{\text{v}D}{\nu} \\ &= \frac{\left(10\ \frac{\text{m}}{\text{s}}\right)(0.0254 \text{ m})}{0.0003\ \frac{\text{m}^2}{\text{s}}} \\ &= 846.7 \quad (850) \end{aligned}$$

The answer is (C).

70. The hydraulic head, or total head, is a measure of the potential of fluid at the measurement point from a datum. The difference in the hydraulic heads between two points in the difference in the hydraulic energy between the points. In this case, the difference in the hydraulic heads of the tanks head is the difference between the elevations of the water surface levels of the tanks.

$$\begin{aligned} \text{Elevation at A} - \text{Elevation at B} &= 10 \text{ ft} + 5 \text{ ft} + 5 \text{ ft} \\ &= 20 \text{ ft} \end{aligned}$$

The answer is (B).

71. Use Newton's second law, given in the Particle Kinetics section of the *NCEES Handbook.*

$$
\begin{aligned}
F &= \frac{d(m\mathrm{v})}{dt} \\
&= \text{mass flow rate} \\
&\qquad \times \text{ velocity change in direction of force} \\
&= \frac{\dot{W}}{g}(\mathrm{v}_2 - \mathrm{v}_1)
\end{aligned}
$$

$\dot{W}$ is the weight of the fluid flow per second. Using the *NCEES Handbook* notation provided in the Fluid Mechanics section, the resultant force in a given direction on water is the rate of change of the momentum of the water.

$$\sum F = Q_2\rho_2\mathrm{v}_2 - Q_1\rho_1\mathrm{v}_1$$

Since water density and flow rate remain unchanged after the impact,

$$
\begin{aligned}
\rho_1 &= \rho_2 = \rho \\
Q_1 &= Q_2 = Q
\end{aligned}
$$

Therefore,

$$F = Q\rho(\mathrm{v}_2 - \mathrm{v}_1)$$

step 1: After impact, the water flow in the 10 in diameter pipe splits into two directions so that the velocity of water in the 10 in diameter pipe in the AB direction is destroyed. Therefore, the impact force equation reduces to

$$F = \left(\frac{w}{g}\right)Q_{10\text{ in}}\mathrm{v}_{10\text{ in}}$$

step 2: Calculate the velocity of the water in the 10 in diameter pipe. The flow rate is

$$Q_{10\text{ in}} = 10\ \text{ft}^3/\text{sec}$$

The cross-sectional area of the pipe is

$$
\begin{aligned}
A_{10\text{ in}} &= \frac{\pi D^2}{4} \\
&= \frac{\pi\left((10\text{ in})\left(\frac{1\text{ ft}}{12\text{ in}}\right)\right)^2}{4} \\
&= 0.55\ \text{ft}^2
\end{aligned}
$$

The water velocity is

$$
\begin{aligned}
\mathrm{v}_{10\text{in}} &= \frac{Q_{10\text{in}}}{A_{10\text{in}}} \\
&= \frac{10\ \frac{\text{ft}^3}{\text{sec}}}{0.55\ \text{ft}^2} \\
&= 18.18\ \text{ft/sec}
\end{aligned}
$$

step 3: Determine the unit weight of water. Use the *NCEES Handbook.* The unit weight of water is given in the Units and Conversion Factors section and in the Thermodynamics: Thermal and Physical Property Tables (at room temperature), as 62.4 lbm/ft^3. The table Properties of Water (English Units) in the Fluid Mechanics section shows that the unit weight of water varies with its temperature. Since no water temperature is provided in the problem statement, use water specific weight, w, of 62.4 lbm/ft^3.

step 4: Determine the force exerted by water on the lock.

$$
\begin{aligned}
F &= \left(\frac{62.4\ \frac{\text{lbm}}{\text{ft}^3}}{32.2\ \frac{\text{lbm-ft}}{\text{lbf-sec}^2}}\right)\left(10\ \frac{\text{ft}^3}{\text{sec}}\right)\left(18.18\ \frac{\text{ft}}{\text{sec}}\right) \\
&= 352.3\ \text{lbf} \quad (350\ \text{lbf})
\end{aligned}
$$

The answer is (B).

72. The maximum height, h, that can be used in siphoning is limited by the absolute pressure needed to avoid separation of dissolved gases. In this case, separation occurs at an absolute pressure of 8 ft of water head, which is less than the atmospheric pressure of 34 ft of water. The difference can be used to raise the pipe elevation above point A. Applying Bernoulli's energy equation to points A and C, the energy at point A = energy at point C + losses between A and C.

$$\frac{p_A}{\gamma} + \frac{\mathrm{v}_A^2}{2g} + z_A = \frac{p_C}{\gamma} + \frac{\mathrm{v}_C^2}{2g} + z_C + h_f + h_{f,\text{fitting}}$$

Express the height as the difference between the heights of point C and point A.

$$h = z_C - z_A$$

In order to maximize the siphoning height, h, the water pressure at point C should be reduced as much as possible without causing separation of the gases from the water. The water velocity at point A is practically zero ($\mathrm{v}_A = 0$), and at point C it is 3 ft/sec. The pipe friction loss and entry loss are given. Therefore, the energy equation reduces to

$$34\text{ ft} + z_A = 8\text{ ft} + \frac{\left(3\ \frac{\text{ft}}{\text{sec}}\right)^2}{(2)\left(32.2\ \frac{\text{ft}}{\text{sec}^2}\right)} + z_C + 15\text{ ft} + 1\text{ ft} + 0.1\text{ ft}$$

$$\begin{aligned} z_A &= z_C - 9.76\text{ ft} \\ h &= z_C - z_A \\ &= 9.76\text{ ft} \quad (9.8\text{ ft}) \end{aligned}$$

The answer is (C).

73. The drag force depends on the air density and the speed. The drag force expression given in the *NCEES Handbook* is

$$F = \frac{C_D \rho v^2 A}{2}$$

The drag coefficient, C_D, depends on the Reynolds number and can be determined using the graph Drag Coefficient for Spheres, Disks, and Cylinders, given in the *NCEES Handbook*, Fluid Mechanics section.

$$\begin{aligned} \text{Re} &= \frac{vD}{\nu} \\ &= \frac{(3.4\text{ m/s})(1\text{ m})}{1.7 \times 10^{-5}\text{ m}^2/\text{s}} \\ &= 200{,}000 \end{aligned}$$

From the graph (using cylinder curve), drag coefficient $C_D = 0.4$.

$$\begin{aligned} F &= \frac{1}{2} \times 0.4 \times (1.12\text{ kg/m}^3) \times (3.4\text{ m/s})^2 \times \pi\left(\frac{1}{2}\text{ m}\right)^2 \\ &= 2.03\text{ N} \quad (2.0\text{ N}) \end{aligned}$$

The answer is (B).

74. The isentropic process is a special case of an adiabatic process in which there is no transfer of heat or matter. See the *NCEES Handbook*, Isentropic Flow Relationships section. Air is compressible, and the figure shows a convergent-divergent nozzle. The problem describes a supersonic flow so that the velocity increases as the flow cross-sectional area increases.

To determine the area at exit point, determine the ratio of the exit area to the throat area (A^*) using the equation.

$$\frac{A}{A^*} = \frac{1}{\text{Ma}}\left[\frac{1 + \frac{1}{2}(k-1)\text{Ma}^2}{\frac{1}{2}(k+1)}\right]^{\frac{(k+1)}{2(k-1)}}$$

The factor k is 1.4 for air. The air velocity at the exit is specified as 5 Ma. Substituting the values, the equation becomes

$$\begin{aligned} \frac{A_{\text{exit}}}{A^*} &= \frac{1}{5}\left[\frac{1 + \frac{1}{2}(1.4-1) \times 5^2}{\frac{1}{2}(1.4+1)}\right]^{\frac{(1.4+1)}{2(1.4-1)}} \\ &= \frac{1}{5}\left[\frac{1 + 0.2 \times 5^2}{1.2}\right]^3 \\ &= 25 \\ A_{\text{exit}} &= 25 \times A_{\text{throat}} \\ &= 25 \times 1.2\text{ in}^2 = 30\text{ in}^2 \end{aligned}$$

The answer is (D).

75. The words *torque* and *torsion* are used interchangeably. From the *NCEES Handbook*, Mechanics of Materials section, by definition the torsion or torque, T, is the shear stress multiplied by the area and lever arm.

step 1: Determine the area.

$$\begin{aligned} A &= \text{Circumferential area of cylinder} \\ &= 2\pi(\text{radius})(\text{length}) \\ &= 2\pi(0.04\text{ m})(0.06\text{ m}) \\ &= 0.01507\text{ m}^2 \end{aligned}$$

step 2: Determine the lever arm.

$$\begin{aligned} \text{LA} &= \text{shaft radius} \\ &= 0.04\text{ m} \end{aligned}$$

step 3: Use the torque equation to determine shear stress, τ. Torque is defined as

$$\begin{aligned} T &= \text{force} \times \text{lever arm} \\ &= (\text{shear stress} \times \text{surface area}) \times \text{shaft radius} \\ &= (\tau)(0.01507\ \text{m}^2)(0.04\ \text{m}) \\ &= (\tau)(6.032 \times 10^{-4}\ \text{m}^3) \end{aligned}$$

It is given that the applied torque is 2 N·m. Therefore,

$$\begin{aligned} T &= (\tau)(6.032 \times 10^{-4}\ \text{m}^3) \\ &= 2\ \text{N·m} \end{aligned}$$

Determine shear stress.

$$\begin{aligned} \tau &= \frac{2\ \text{N·m}}{6.032 \times 10^{-4}\ \text{m}^3} \\ &= 3315.65\ \text{N/m}^2 \end{aligned}$$

step 4: Use the Newtonian fluid film formula to determine viscosity. Since the shear stress at the interface is caused by the viscosity of the fluid, use the *NCEES Handbook*, Stress, Pressure, and Viscosity section. The shear stress and viscosity are related.

$$\tau = \mu \frac{d\text{v}}{dy}$$

For Newtonian fluid film, simplify the above equation.

$$\begin{aligned} \tau &= \mu \frac{d\text{v}}{dy} \\ &= \mu \frac{\text{v}}{\delta} \\ &= \mu \times \frac{\text{velocity of shaft on film}}{\text{thickness of fluid film}} \\ &= \mu \times \frac{r \times \omega}{0.001\ \text{m}} \end{aligned}$$

step 5: The shaft is rotating with an angular velocity, ω. Convert the angular velocity to linear velocity.

$$\text{v} = r \times \omega = (\text{shaft radius})(\text{angular velocity})$$

$$\begin{aligned} \tau &= \mu \frac{(0.04\ \text{m})\left(1000\ \frac{\text{rad}}{\text{s}}\right)}{0.001\ \text{m}} \\ &= 40{,}000\ \mu \\ &= 3315.65\ \text{N/m}^2 \end{aligned}$$

The viscosity is

$$\begin{aligned} \mu &= \frac{3315.65\ \frac{\text{N}}{\text{m}^2}}{40{,}000\ \frac{\text{rad}}{\text{s}}} \\ &= 0.08289\ \text{N·s/m}^2 \quad (0.083\ \text{N·s/m}^2) \end{aligned}$$

The answer is (B).

76. A system head curve is defined as the relation between flow and head required in a fixed hydraulic network. The equation is written on the total system, excluding the pump. The pump must overcome pressure, velocity, elevation, and friction both upstream and downstream to lift the water from reservoir 1 to reservoir 2.

The head demand for the system is the difference between the energy levels at points 1 and 2. For energy equation, see *NCEES Handbook*, Principles of One-Dimensional Fluid Flow: Energy Equation section.

The energy equation for steady incompressible flow with no shaft device is

$$\frac{p_1}{\gamma} + z_1 + \frac{\text{v}_1^2}{2g} = \frac{p_2}{\gamma} + z_2 + \frac{\text{v}_2^2}{2g} + h_f$$

$$\frac{p_2 - p_1}{\gamma} + \frac{\text{v}_2^2 - \text{v}_1^2}{2g} + (z_2 - z_1) + h_f = 0$$

When a pump is added to lift the water elevation to point 2, the system head demand is

$$H_{\text{system}} = \frac{p_2 - p_1}{\gamma} + \frac{\text{v}_2^2 - \text{v}_1^2}{2g} + (z_2 - z_1) + h_f$$

step 1: To solve the problem, use system properties to simplify the general equation.

$$\begin{aligned} p_2 &= p_1 \\ \text{v}_2 &= \text{v}_1 = 0 \\ z_2 - z_1 &= 100\ \text{ft} \end{aligned}$$

$$H_{\text{system}} = 100\ \text{ft} + h_f$$

step 2: The friction coefficient f is given as 0.0198. The head loss due to friction is

$$h_f = f \frac{l}{d} \frac{\text{v}^2}{2g}$$

The pipe diameter and area of cross section are

$$d = 6 \text{ in} = 0.5 \text{ ft}$$
$$A = \frac{\pi d^2}{4} = \frac{\pi(0.5 \text{ ft})^2}{4} = 0.196 \text{ ft}^2$$

$$h_f = f\frac{l}{d}\frac{v^2}{2g} = f\frac{l}{d}\frac{1}{2g}\left(\frac{Q}{A}\right)^2$$
$$= 0.0198\left(\frac{1000 \text{ ft}}{0.5 \text{ ft}}\right)\frac{1}{2(32.2 \text{ ft/sec}^2)}\left(\frac{Q}{0.196 \text{ ft}^2}\right)^2$$
$$= 16Q^2$$

Therefore, the system head demand is

$$H_{\text{system}} = 100 + 16Q^2$$

The answer is 100 + 16Q².

77. Convert gallons of water into lbf. See the *NCEES Handbook*, Units and Conversion Factors section.

$$1 \text{ gal water} = 8.3453 \text{ lbf}$$

step 1: See the *NCEES Handbook*, Pump Power Equation section, and convert water discharge to power.

$$\dot{W}_1 = \frac{Q\gamma h}{\eta}$$
$$= \left(\frac{670 \text{ gal}}{1 \text{ min}}\right)\left(\frac{8.3453 \text{ lbf}}{1 \text{ gal}}\right)\left(\frac{65 \text{ ft}}{0.7}\right) \times \left(\frac{1 \text{ hp}}{33000 \text{ ft-lbf/min}}\right)$$
$$= 15.73 \text{ hp}$$

See the *NCEES Handbook*, Performance of Components section, for scaling laws for pumps. The pump's power requirement depends on the pump's rotor speed.

step 2: For rotational speed scaling from pump 1 to pump 2, apply the relation

$$\left(\frac{H}{N^2D^2}\right)_1 = \left(\frac{H}{N^2D^2}\right)_2$$
$$N_2^2 = N_1^2\left(\frac{D_1^2}{D_2^2}\right)\left(\frac{H_2}{H_1}\right)$$

Apply the pump properties.

$$\text{impeller diameter, } D_2 = D_1$$
$$\text{fluid density, } \rho_2 = \rho_1$$
$$\text{head, } H_1 = 65 \text{ ft}$$
$$H_2 = 100 \text{ ft}$$
$$\text{rotational speed, } N_1 = 1500 \text{ rpm}$$

$$N_2^2 = (1500)^2 \times (1) \times \left(\frac{100 \text{ ft}}{65 \text{ ft}}\right)$$
$$N_2 = (1500)(1.24) = 1860.5 \text{ rpm}$$

step 3: Apply the power scaling relation.

$$\left(\frac{\dot{W}}{\rho N^3 D^5}\right)_2 = \left(\frac{\dot{W}}{\rho N^3 D^5}\right)_1$$
$$\dot{W}_2 = \dot{W}_1\left(\frac{N_2}{N_1}\right)^3 = 15.73 \text{ hp} \times \left(\frac{1860.5 \text{ rpm}}{1500 \text{ rpm}}\right)^3$$
$$= 30.02 \text{ hp} \quad (30 \text{ hp})$$

The answer is (C).

78. Use the *NCEES Handbook*, The Impulse-Momentum Principle: Deflectors and Blades section, for a moving blade. The force exerted by the jet in the x-direction is

$$-F_x = -Q\rho(v_1 - v)(1 - \cos\alpha)$$
$$F_x = Q\rho(v_1 - v)(1 - \cos\alpha)$$

The above equation is based on the principle

$$\text{force} = (\text{mass})(\text{acceleration})$$
$$= (\text{mass flow per second}) \times \left(\text{change of velocity in the direction of force}\right)$$

Calculate the relative velocity with the vane moving.

$$v_{\text{rel}} = v_1 - v$$
$$= 300 \frac{\text{ft}}{\text{sec}} - 100 \frac{\text{ft}}{\text{sec}}$$
$$= 200 \text{ ft/sec}$$

The discharge rate from the pipe is

$$\begin{aligned} Q &= A\text{v} \\ &= \left(\frac{\pi d^2}{4}\right)\left(300\ \frac{\text{ft}}{\text{sec}}\right) \\ &= \left(\frac{\pi}{4}(1\ \text{in})\left(\frac{1\ \text{ft}}{12\ \text{in}}\right)^2\right)\left(300\ \frac{\text{ft}}{\text{sec}}\right) \\ &= 1.64\ \text{ft}^3/\text{sec} \end{aligned}$$

If the force is to be measured in pounds, the mass must be in slugs.

Mass flow rate per second is

$$\begin{aligned} Q(\rho) &= \left(1.64\ \frac{\text{ft}^3}{\text{sec}}\right)\left(\frac{62.4\ \frac{\text{lbm}}{\text{ft}^3}}{32.2\ \frac{\text{lbm}}{\text{slug}}}\right) \\ &= 3.28\ \text{slugs/sec} \end{aligned}$$

The force in the x-direction is

$$\begin{aligned} F_x &= \left(3.28\ \frac{\text{slugs}}{\text{sec}}\right)\left(200\ \frac{\text{ft}}{\text{sec}}\right)(1 - \cos 60^\circ) \\ &= 328\ \text{lbf} \end{aligned}$$

To compute the power of the turbine in horsepower units, use the factor in *NCEES Handbook*, Units and Conversion Factors section.

$$\begin{aligned} W &= F_x\text{v} \\ &= (328\ \text{lbf})\left(100\ \frac{\text{ft}}{\text{sec}}\right) \\ &= \left(32{,}800\ \frac{\text{ft-lbf}}{\text{sec}}\right)\left(\frac{1\ \text{hp}}{550\ \frac{\text{ft-lbf}}{\text{sec}}}\right) \\ &= 59.64\ \text{hp} \quad (60\ \text{hp}) \end{aligned}$$

The answer is (D).

79. A single component system in thermodynamics has two types of properties: intensive and extensive. The intensive properties are independent of mass; and the extensive properties are proportional to the mass of the system.

For example, the pressure and temperature are independent of the mass. Therefore, they are considered as the intensive properties.

The ratio of any extensive property of the system to the mass of the system is called the specific property. For example, specific volume is defined as volume per unit mass. While volume depends on the mass of the system, the specific volume is independent of the mass. Therefore, it is an intensive property of the system. Thus, options B, C, and D are intensive properties.

The answer is (A).

80. See the *NCEES Handbook*, Thermodynamics section.

step 1: Apply the first law of thermodynamics, which states that the energy is conserved in a heat system and that, if no mass crosses the boundary, then energy change is

$$Q - W = \Delta U + \Delta KE + \Delta PE$$

Since the changes in kinetic and potential energy are zero in this case, heat energy change is

$$Q - W = \Delta U$$

Since the container is rigid, no work is done. Therefore, heat energy is

$$Q = \Delta U = m(u_2 - u_1)$$

step 2: From the superheated water pressure table for $p = 0.2$ MPa and 300°C,

$$u_1 = 2808.6\ \text{kJ/kg}$$
$$v_1 = 1.3162\ \text{m}^3/\text{kg}$$

step 3: For state 2, because the container is rigid, specific volume

$$\begin{aligned} v_2 &= v_1 \\ &= 1.3162\ \text{m}^3/\text{kg} \end{aligned}$$

step 4: Determine the values of internal energy and temperature for the required pressure of 0.4 MPa. For this, use the superheated water pressure table (for constant pressure). Select constant pressure, $p = 0.4$ MPa. The table does not have the values pertaining to the specific volume of 1.3162 m³/kg. Therefore, select two rows that bound the specific volume value and then interpolate. The two rows are

T (°C)	v (m³/kg)	u (kJ/kg)
800	1.2372	3662.4
900	1.3529	3853.9

By interpolation, corresponding with the known specific volume at state 2,

$$v_2 = 1.3162 \text{ m}^3/\text{kg}$$
$$u_2 = 3793.2 \text{ kJ/kg}$$
$$T = 868°\text{C}$$

step 5: Apply the heat energy equation for rigid containers.

$$\begin{aligned}\Delta U &= m(u_2 - u_1) \\ &= (2 \text{ kg})(3793.2 \text{ kJ/kg} - 2808.6 \text{ kJ/kg}) \\ &= 1969.1 \text{ kJ} \quad (1970 \text{ kJ})\end{aligned}$$

The answer is (A).

81. See the *NCEES Handbook*, Thermodynamics section.

step 1: Draw the T-v diagram for the compressor as shown. At point 1, vapor is superheated and under constant pressure from point 1 to 2. At point 2, the vapor is saturated. Between points 2 and 3, the vapor starts phasing from saturated vapor to saturated liquid. The process is complete at point 3.

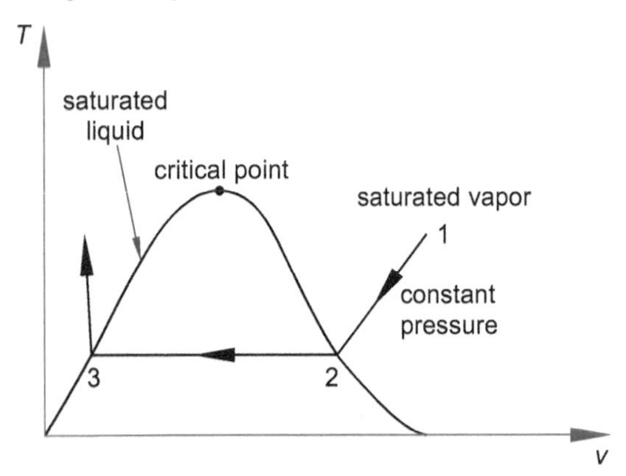

step 2: Apply the first law of thermodynamics. The heat transfer needed to decrease the temperature of a flowing mass, $\dot{m}$, from point 2 to point 3 under constant pressure is

$$\begin{aligned}Q_{\text{in}} &= \dot{m}(h_2 - h_3) \\ &= \dot{m}h_{\text{fg}}\end{aligned}$$

step 3: For a constant pressure of 1 MPa, see the steam tables, and read the enthalpy change value, h_{fg}. The corresponding energy loss is

$$\begin{aligned}Q_{\text{in}} &= \dot{m}h_{\text{fg}} \\ &= \left(10 \ \frac{\text{kg}}{\text{s}}\right)\left(2015 \ \frac{\text{kJ}}{\text{kg}}\right) \\ &= 20{,}150 \text{ kJ/s}\end{aligned}$$

step 4: The heat loss is used to heat the flowing water mass, $\dot{m}_w$.

$$Q_{\text{out}} = \dot{m}_w c_p \Delta T$$

See the Thermal and Physical Property Tables in the *NCEES Handbook* and read the specific heat value of water at constant pressure. Use temperature values in kelvins.

$$\begin{aligned}Q_{\text{out}} &= \dot{m}_w c_p(303\text{K} - 293\text{K}) \\ &= \dot{m}_w\left(4.18 \ \frac{\text{kJ}}{\text{kg}\cdot\text{K}}\right)(10\text{K}) \\ &= \dot{m}_w\left(41.8 \ \frac{\text{kJ}}{\text{kg}}\right)\end{aligned}$$

step 5: Equate the input and output energy levels.

$$\begin{aligned}Q_{\text{out}} &= Q_{\text{in}} \\ (41.8 \text{ kJ/kg})\dot{m}_w &= 20{,}150 \ \frac{\text{kJ}}{\text{s}} \\ \dot{m}_w &= \frac{20{,}150 \ \frac{\text{kJ}}{\text{s}}}{41.8 \ \frac{\text{kJ}}{\text{kg}}} \\ &= 482 \text{ kg/s}\end{aligned}$$

The answer is 482 kg/s.

82. See the *NCEES Handbook*, Thermodynamics section, Mollier (h, s) Diagram for Steam.

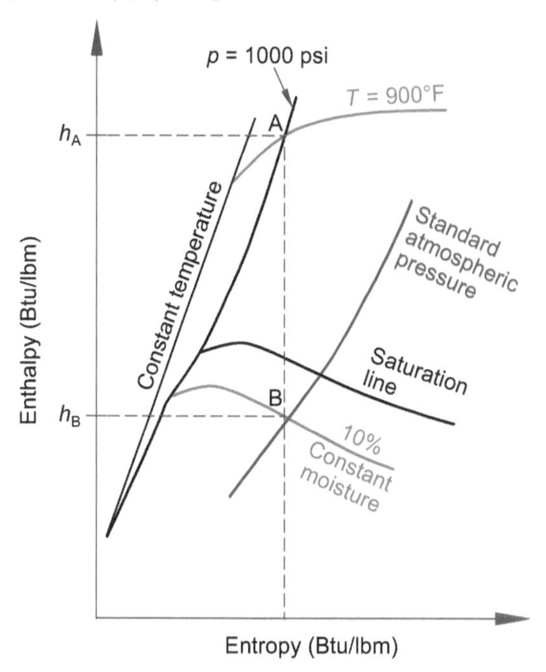

step 1: Locate the initial pressure of 1000 psia and 900°F on the diagram as point A. Enthalpy, $h_A = 1450$ Btu/lbm of steam.

step 2: It is given that the steam expands isentropically. Therefore, draw a straight line down to the standard atmospheric pressure line, shown as point B. The enthalpy is $h_B = 1051$ Btu/lbm of steam.

step 3: Determine the drop in enthalpy for 1 lbm of steam.

$$\begin{aligned}\Delta h &= h_A - h_B \\ &= 1\ \text{lbm}\left(1450\ \frac{\text{Btu}}{\text{lbm}} - 1051\ \frac{\text{Btu}}{\text{lbm}}\right) \\ &= 399\ \text{Btu}\end{aligned}$$

step 4: At point B, read constant moisture. It is 10%.

The answer is 399 Btu drop in enthalpy, and 10% constant moisture.

83. See the *NCEES Handbook*, Thermodynamics specifications. The knowledge area is part of the specification, but no information is provided in the *NCEES Handbook*.

The device to keep the engine speed more or less uniform at all load conditions is called a governor. Therefore, option A is incorrect.

The device that mixes air and fuel in an appropriate air-fuel ratio for combustion is called a carburetor. Therefore, option B is a correct choice.

The device for firing the explosive mixture in an internal combustion engine is called a spark plug. Therefore, option C is incorrect.

The piston transfers power to a crankshaft via a piston rod and/or connecting rod. The device used to change phase in a cycle is called a compressor or an expansion chamber. Therefore, option D is incorrect.

The device to transfer force from expanding gas in the cylinder is called a piston. Therefore, option E is correct.

Options A, C, and D are incorrect, and options B and E are correct.

The answer is carburetor (B) and piston (E).

84. See the *NCEES Handbook*, Thermodynamics section. If a heat cycle takes in heat Q_H at a high temperature T_H and rejects heat Q_L at a low temperature, T_L, then its efficiency is given by

$$\eta = \frac{W}{Q_H} = \frac{Q_H - Q_L}{Q_H}$$

step 1: Draw a line system diagram as shown here and determine work, W.

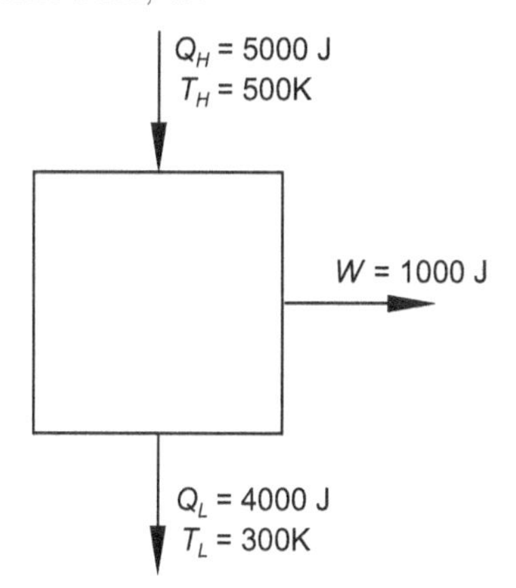

step 2: Determine the efficiency of the system using the formula

$$\begin{aligned}\eta &= \frac{W}{Q_H} = \frac{Q_H - Q_L}{Q_H} \\ &= \frac{5000\ \text{J} - 4000\ \text{J}}{5000\ \text{J}} \times 100\% \\ &= 20\%\end{aligned}$$

step 3: The Carnot cycle is considered to be the most efficient cycle, and its efficiency is

$$\begin{aligned}\eta_c &= \frac{T_H - T_L}{T_H} \\ &= \frac{500\text{K} - 300\text{K}}{500\text{K}} \times 100\% \\ &= 40\%\end{aligned}$$

step 4: Determine the ratio of efficiencies.

$$\frac{\eta}{\eta_c} = \frac{20\%}{40\%} \times 100\% = 50\%$$

The correct answer is 0.5.

85. See the *NCEES Handbook*, Thermodynamics section. The cycle described in the problem is a refrigeration cycle. Pressure versus enthalpy curves and property tables for R-410A are given in the *NCEES Handbook*.

step 1: Use the *p-h* diagram and the given pressures and draw the cycle as shown.

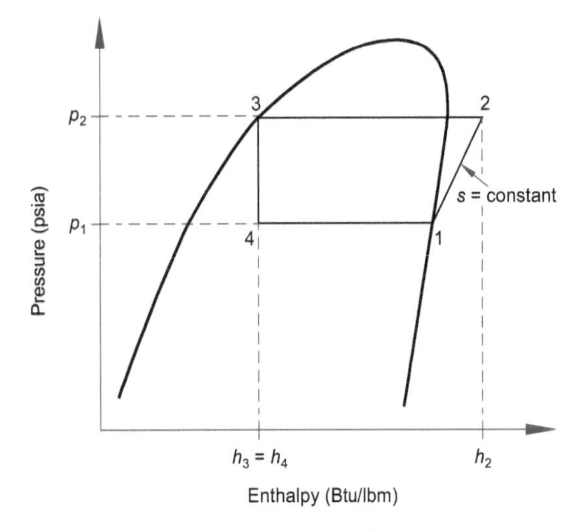

step 2: Use the table and read off the following values at the saturated liquid-vapor dome. For 30 psia saturated vapor,

$$h_1 = 113.88 \ \frac{\text{Btu}}{\text{lbm}}$$

$$s_1 = 0.26711 \ \frac{\text{Btu}}{\text{lbm}}$$

For 200 psia, constant pressure saturated liquid,

$$\begin{aligned}h_3 &= h_4 \\ &= 37.22 \ \frac{\text{Btu}}{\text{lbm}}\end{aligned}$$

step 3: At point 3, entropy $s_3 = s_1 = 0.2675$ Btu/lbm. Use the *p-h* diagram and interpolate between the $s = 0.26$ and $s = 0.28$ isentropic lines to obtain the entropy value at point 3. Read the enthalpy value at point 3 as close as possible. For 200 psia and $s = 0.2675$ Btu/lbm,

$$h_2 \approx 136 \ \frac{\text{Btu}}{\text{lbm}}$$

step 4: Determine the heat rejected in the condenser.

$$\begin{aligned}Q_{\text{condenser}} &= h_2 - h_3 \\ &= 136 \ \frac{\text{Btu}}{\text{lbm}} - 37.22 \ \frac{\text{Btu}}{\text{lbm}} \\ &= 98.78 \ \frac{\text{Btu}}{\text{lbm}} \quad (100 \text{ Btu/lbm})\end{aligned}$$

The correct answer is 100 Btu/lbm.

86. See the *NCEES Handbook*, Chemistry and Thermodynamics sections. The change in entropy in a nonreacting gas mixture depends on the mixing process. In this case, both temperature and pressure are kept constant, and volume is allowed to vary. For each gas in the mixture, the change in specific entropy is

$$\Delta s = R \ln\left(\frac{V_{\text{mixture}}}{V_i}\right)$$

step 1: Determine the volume of each gas.

$$\text{Molecular mass of oxygen} = 32 \text{ kg/m}^3$$

$$\text{Molecular mass of nitrogen} = 28 \text{ kg/m}^3$$

$$\begin{aligned}\text{Initial volume of oxygen} &= \frac{8 \text{ kg}}{32 \ \frac{\text{kg}}{\text{m}^3}} \\ &= 0.25 \text{ m}^3\end{aligned}$$

$$\begin{aligned}\text{Initial volume of nitrogen} &= \frac{7 \text{ kg}}{28 \ \frac{\text{kg}}{\text{m}^3}} \\ &= 0.25 \text{ m}^3\end{aligned}$$

$$\begin{aligned}\text{Volume of mixture} &= 0.25 \text{ m}^3 + 0.25 \text{ m}^3 \\ &= 0.50 \text{ m}^3\end{aligned}$$

step 2: Determine the constant of each gas. The universal gas constant, $\bar{R}$, is 8.314. The gas constant is specific to each gas and is given by the expression

$$R_i = \frac{\overline{R}}{\text{mol mass}}$$

For oxygen and nitrogen, the gas constants are

$$R_{\text{oxygen}} = \frac{8.314}{32} = 0.26$$
$$R_{\text{nitrogen}} = \frac{8.314}{28} = 0.297$$

step 3: Determine the change in specific entropy in oxygen.

$$\begin{aligned}\Delta s &= R_{\text{oxygen}} \ln\left(\frac{V_{\text{mixture}}}{V_{\text{oxygen}}}\right) \\ &= (0.26)\ln\left(\frac{0.5 \text{ m}^3}{0.25 \text{ m}^3}\right) \\ &= (0.26)\ln 2 \\ &= 0.26(0.6931) \\ &= 0.18 \text{ kJ/K}\end{aligned}$$

The answer is 0.18 kJ/K.

87. See the *NCEES Handbook*, Thermodynamics section. Use the psychrometric chart and read off the results as shown. The chart is a plot of atmospheric air properties as a function of dry-bulb temperature (t_{db}).

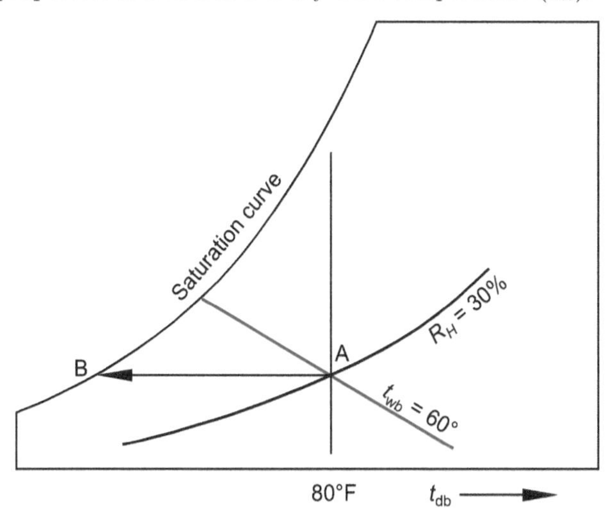

step 1: The air condition is 80°F dry-bulb temperature, t_{db}, and 60° F wet-bulb temperature, t_{wb}. Locate the condition on the chart at point A.

step 2: Read the relative humidity along the relative humidity curve. It is 30%.

step 3: From point A, draw a horizontal line meeting the saturation curve at point B as shown. The dew-point temperature, t_{dp}, as read on the saturation curve is 46°F.

The answer is relative humidity = 30%, dew-point temperature = 46°F.

88. See the *NCEES Handbook*, Thermodynamics section. The stoichiometric equation for combustion of methane in air given in the *Handbook* is

$$CH_4 + 2O_2 + 2(3.76)N_2 \rightarrow CO_2 + 2\ H_2O + 7.52\ N_2$$

The equation shows that it takes two volumes of oxygen to burn one volume of methane. To calculate mass products of combustion, use the periodic table of elements given in the Chemistry section of the *NCEES Handbook*. In terms of moles, the above equation becomes

$$\left(\begin{array}{c}1 \text{ mol } CH_4 + 2 \text{ mol } O_2 \\ + 7.52 \text{ mol } N_2\end{array}\right) = \left(\begin{array}{c}2 \text{ mol } CO_2 + 2 \text{ mol } H_2O \\ + 7.52 \text{ mol } N_2\end{array}\right)$$
$$\left(\begin{array}{c}16 \text{ kg} + 2(32 \text{ kg}) \\ + 7.52(28 \text{ kg})\end{array}\right) = \left(\begin{array}{c}2(44 \text{ kg}) + 2(18 \text{ kg}) \\ + 7.52(28 \text{ kg})\end{array}\right)$$

To determine the products of combustion for 1 kg methane, divide both sides of the equation by 16.

$$\left(\begin{array}{c}\frac{16 \text{ kg}}{16} + \frac{2(32 \text{ kg})}{16} \\ + \frac{7.52(28 \text{ kg})}{16}\end{array}\right) = \left(\begin{array}{c}\frac{2(44 \text{ kg})}{16} + \frac{2(18 \text{ kg})}{16} \\ + \frac{7.52(28 \text{ kg})}{16}\end{array}\right)$$
$$\begin{aligned}1 \text{ kg} + 4 \text{ kg} + 13.16 \text{ kg} &= 5.5 \text{ kg} + 2.25 \text{ kg} + 13.16 \text{ kg} \\ &= 20.91 \text{ kg} \quad (21 \text{ kg})\end{aligned}$$

The answer is 21 kg.

89. See the *NCEES Handbook*, Heat Transfer section. Newton's law of cooling states that the heat transferred from a hot body to a cold body is directly proportional to (1) the surface area of the heat flow and (2) the difference of temperatures between the two bodies. The formula for the Newton's cooling law is

$$\dot{Q} = hA(T_w - T_\infty)$$

Therefore, statement A is incorrect.

Fourier's law concerns the heat conduction. In conduction, heat transfers from one particle of the body to the other. The term x refers to the thickness of the body through which heat flows. Therefore, statement B is incorrect.

Radiation is a process of heat transfer from a hot body to cold body, in a straight line, without affecting the intervening medium. Therefore, statement C is incorrect.

Thermal conductivity of solid metals decreases with increase in temperature; and thermal resistivity of solid metals increases with increase in temperature. Therefore, statement D is correct.

The answer is (D).

90. Use Fourier's law of conduction. See *NCEES Handbook*, Heat Transfer section, for the rate of heat transfer through a plane wall.

$$\begin{aligned}\dot{Q} &= kA\frac{dT}{dx}\\ &= (1\ \text{W/m·k})(2.5\ \text{m})(6\ \text{m})\frac{\left(20^\circ\text{C}-(-20^\circ\text{C})\right)}{0.15\ \text{m}}\\ &= 4000\ \text{W}\\ &= 4\ \text{kW}\end{aligned}$$

The answer is 4 kW.

91. See the *NCEES Handbook*, Heat Transfer: Convection section.

Conduction is a process wherein heat transfers from one particle of the body to another in the direction of fall of temperature, while the particles remain in fixed position relative to each other. Therefore, option A matches with statement 1.

Convection is the process wherein heat flows from one particle of the body to another in the direction of fall of temperature, while the particles move relative to each other. Therefore, option B matches with statement 2.

Stefan-Boltzmann law defines the change in heat flow rate as the difference in the fourth powers of the absolute temperatures of the object and of its environment. Therefore, option E matches with statement 3.

The answer is drag option A to statement 1, drag option B to statement 2, and drag option E to statement 3.

92. See the *NCEES Handbook*, Heat Transfer section. The radiation emitted by a body is given by

$$\dot{Q} = \varepsilon\sigma A T^4$$

step 1: A gray body is one that has emissivity and reflectivity, so that

$$\begin{aligned}\text{emissivity factor}, \varepsilon &= 1 - \text{reflectivity factor}\\ &= 1 - 0.15\\ &= 0.85\end{aligned}$$

step 2: Determine variables.

$$\begin{aligned}\text{Stefan-Boltzmann constant}, \sigma &= 5.67\times10^{-8}\ \text{W/m}^2\cdot\text{K}^4\\ \text{Surface area of cylinder},\ A &= \pi DL\\ &= \pi(0.02\ \text{m})(1.98\ \text{m})\\ &= 0.1244\ \text{m}^2\\ \text{Element temperature},\ T_1 &= 1000^\circ\text{C}+273 = 1273\text{K}\\ \text{Wall temperature},\ T &= 500^\circ\text{C}+273 = 773\text{K}\end{aligned}$$

step 3: Apply the data to the radiation equation.

$$\begin{aligned}\dot{Q} &= \varepsilon\sigma A T^4\\ &= (0.85)(5.67\times10^{-8}\ \text{W/m}^2\cdot\text{K}^4)(0.1244\ \text{m}^2)\\ &\quad\times\left((1273\text{K})^4-(773\text{K})^4\right)\left(\frac{1\ \text{kW}}{1000\ \text{J/s}}\right)\\ &= 13.60\ \text{kW}\quad (14\ \text{kW})\end{aligned}$$

The answer is (B).

93. See the *NCEES Handbook*, Heat Transfer section. The Biot number is used in transient conduction problems in which a solid body experiences a sudden change in its thermal environment. It is defined as

$$B_i = \frac{hV}{kA_s}$$

See the *NCEES Handbook*, Mathematics section. For a sphere,

$$\begin{aligned}V &= \frac{\pi d^3}{6}\\ A_s &= \pi d^2\\ \frac{V}{A_s} &= \frac{\frac{\pi d^3}{6}}{\pi d^2} = \frac{d}{6} = \frac{9\ \text{mm}}{6}\\ &= 1.5\ \text{mm}\\ &= 1.5\times10^{-3}\ \text{m}\end{aligned}$$

The following information is given.

$$\begin{aligned}h &= 400\ \text{W/m}^2\cdot\text{K}\\ k &= 20\ \text{W/m}\cdot\text{K}\end{aligned}$$

Calculate the Biot number as

$$\begin{aligned} B_i &= \frac{hV}{kA_s} \\ &= \frac{(400\ \text{W/m}^2\text{·K})(1.5\times 10^{-3}\ \text{m})}{20\ \text{W/m·K}} \\ &= 0.03 \end{aligned}$$

The answer is (B).

94. See the *NCEES Handbook*, Heat Transfer: Heat Exchangers section. The pressure, temperature, and specific heat are thermodynamics properties. while terms such as heat capacity, heat transfer, heat exchange, and specific heat are used in thermodynamics, but "heat" is a vague term and not a thermodynamics property.

The answer is (A).

95. Use the *NCEES Handbook*, Heat Transfer. The problem involves a pipe surrounded by insulation. The heat loss is through conductivity. Therefore, compute the rate of heat transfer by conduction through a cylindrical wall. It is expressed as

$$\dot{Q} = \frac{2\pi kL(T_1 - \text{T}_2)}{\ln\left(\frac{r_2}{r_1}\right)}$$

step 1: From the data, the heat loss parameters are

$$\begin{aligned} k &= 0.2\ \text{W/m·K} \\ L &= 1\ \text{m} \\ T_1 &= 200°\text{C} + 273 = 473\text{K} \\ T_2 &= \ 85°\text{C} + 273 = 358\text{K} \\ r_1 &= \frac{150\ \text{mm}}{2} = 75\ \text{mm} \\ r_2 &= 75\ \text{mm} + 50\ \text{mm} = 125\ \text{mm} \end{aligned}$$

step 2: Apply the data to the heat loss formula.

$$\begin{aligned} \dot{Q} &= \frac{2\pi kL(T_1 - T_2)}{\ln\left(\frac{r_2}{r_1}\right)} \\ &= \frac{\begin{array}{l} 2\pi(0.2\ \text{W/m·K})(1\ \text{m}) \\ \quad \times(473\text{K} - 358\text{K})(60\ \text{sec/min}) \end{array}}{\ln\left(\frac{125\ \text{mm}}{75\ \text{mm}}\right)} \\ &= 16{,}974\ \text{J} \quad (17{,}000\ \text{J}) \end{aligned}$$

The answer is (B).

96. See the *NCEES Handbook*, Instrumentation, Measurement, and Control, and Fluid Mechanics sections. As stated in the Fluid Mechanics section, absolute pressure is equal to the atmospheric pressure minus the vacuum gauge pressure reading.

The vacuum gauge pressure is the difference between the local atmospheric pressure and the absolute pressure. Therefore, option A is a correct statement.

As stated in the *NCEES Handbook*, the pressure sensors are called pressure transducers, pressure transmitters, pressure senders, pressure indicators, piezometers, and manometers. The sensors are typically based on measuring the strain on a thin membrane due to an applied pressure. Based on their design, pressure sensors are used to measure the absolute, gauge, or differential pressure. Therefore, option B is a correct statement.

As stated in the *NCEES Handbook*, the transducer sensitivity is defined as the ratio of change in electrical signal magnitude to the change in magnitude of the physical parameter being measured. Therefore, option C is a correct statement.

The Wheatstone bridge configuration is shown in the *NCEES Handbook*, Instrumentation, Measurement, and Control section. For the bridge to be balanced, all four of its resistances need not be equal. It is sufficient that

$$\frac{R_1}{R_3} = \frac{R_2}{R_4}$$

Therefore, the statement in option D is incorrect.

The answer is (D).

97. The block diagrams are used to study the automatic control systems. In this, functioning of a system is explained by the interconnected block. Each block is labeled and represents a specific function. The blocks are connected to other blocks by lines with arrow marks that indicate the sequence of the events. Every block has an input and an output. Several blocks connected in series represent a control system.

The illustration shows a basic negative feedback control model block diagram.

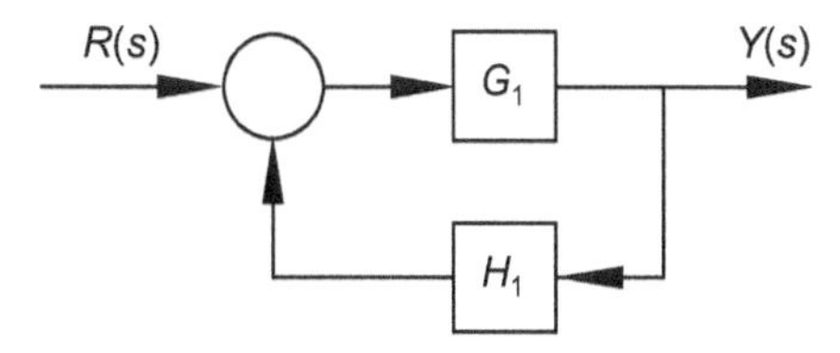

In this diagram, G_1 represents a controller or compensator function, and H_1 represents a measurement dynamic. The function $R(s)$ is the input, and $Y(s)$ is the output function.

$$Y(s) = \frac{G_1}{1 + G_1H_1}R(s)$$

This is a simplified version of the formula given in the *NCEES Handbook*, Instrumentation, Measurement, and Control section.

In this problem, two control systems, G_1 and G_2, represent the basic negative feedback control model shown above, and their output can be computed using the above formula. Therefore, the system can be represented as the sum of two blocks.

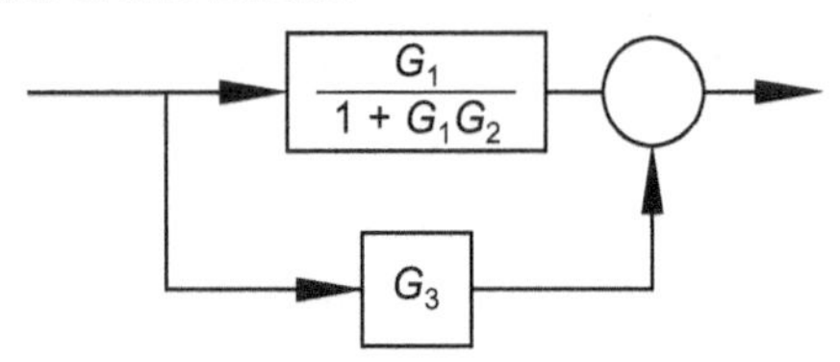

The block G_3 provides positive feedback to the system. Therefore, the output of the above system can be computed by adding the block functions as follows.

$$Y(s) = \frac{G_1}{1+G_1G_2} + G_3$$

The equation is in the form required in the problem statement. Therefore,

$$\begin{aligned} X &= G_1 \\ Y &= G_1G_2 \\ Z &= G_3 \end{aligned}$$

The answer is (B).

98. See the *NCEES Handbook*, Dynamics. The term resonance speed pertains to the natural free frequency of the motor, which is defined as

$$\omega_n = \sqrt{\frac{k}{m}}$$

step 1: Using ft-lbf units, the spring stiffness is

$$\begin{aligned} k &= 4(1000 \text{ lbf/in})(12 \text{ in/1 ft}) \\ &= 48{,}000 \text{ lbf/ft} \end{aligned}$$

step 2: Convert the motor weight to mass.

$$\begin{aligned} m &= \frac{w}{g} = \frac{400 \text{ lbf}}{32.2 \text{ ft/sec}^2} \\ &= 12.42 \ \frac{\text{lbf-sec}^2}{\text{ft}} \end{aligned}$$

step 3: Apply the formula and convert frequency from rad/sec to rpm.

$$\begin{aligned} \omega_n &= \sqrt{\frac{48{,}000 \ \frac{\text{lbf}}{\text{ft}}}{12.42 \ \frac{\text{lbf-sec}^2}{\text{ft}}}} \\ &= 62.16 \ \frac{\text{rad}}{\text{sec}} \\ &= 62.16 \ \frac{\text{rad}}{\text{sec}}\left(\frac{60 \text{ sec}}{1\text{min}}\right)\left(\frac{1 \text{ rev}}{2\pi \text{ rad}}\right) \\ &= 594 \text{ rpm} \quad (590 \text{ rpm}) \end{aligned}$$

The answer is (B).

99. See the *NCEES Handbook*, Instrumentation, Measurement, and Control section. Measurement accuracy is defined as closeness of agreement between a measured quantity value and a true quantity value of a measurement. When reporting measurement results, it is necessary to provide an associated uncertainty so that those who use it may assess its reliability. One method to assess the uncertainty is to use the Kline-McClintock general equation.

$$w_R = \sqrt{\left(w_1\frac{\partial f}{\partial x_1}\right)^2 + \left(w_2\frac{\partial f}{\partial x_2}\right)^2 + \cdots + \left(w_n\frac{\partial f}{\partial x_n}\right)^2}$$

In this case, the uncertainties in the measurements of distance and time are known. The uncertainty in measurement of speed needs to be computed.

step 1: Determine the average velocity of the particle.

$$\text{v} = \frac{100 \text{ m}}{25 \text{ s}} = 4 \ \frac{\text{m}}{\text{s}}$$

The uncertainty in velocity measurement is expressed as

$$\text{v} = 4 \text{ m} \pm dv \ \frac{\text{m}}{\text{s}}$$

step 2: Apply the Kline-McClintock equation to the problem.

$$\begin{aligned}\frac{\delta \text{v}}{t} &= \sqrt{\left(\frac{\delta d}{d}\right)^2 + \left(\frac{\delta t}{t}\right)^2} \\ &= \sqrt{\left(\frac{1 \text{ m}}{100 \text{ m}}\right)^2 + \left(\frac{3 \text{ s}}{25 \text{ s}}\right)^2} \\ &= \sqrt{(0.01)^2 + (0.12)^2} \\ &= \sqrt{0.0154} \\ &= 0.1204\end{aligned}$$

step 3: Determine the uncertainty in velocity.

$$\begin{aligned}\delta \text{v} &= 0.1204\text{v} \\ &=(0.1204 \text{ m})(4 \text{ m/s}) \\ &= 0.482 \ \frac{\text{m}}{\text{s}} \quad (0.5 \text{ m/s})\end{aligned}$$

step 4: The measured speed is

$$\text{v} = 4 \ \frac{\text{m}}{\text{s}} \pm 0.5 \ \frac{\text{m}}{\text{s}}$$

The answer is 4 m/s ± 0.5 m/s.

100. A basic negative feedback control model block diagram is shown.

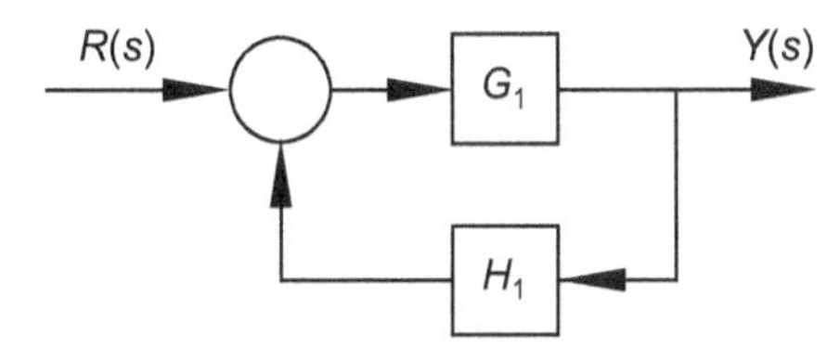

This system's output function is given by

$$Y(s) = \frac{G_1}{1 + G_1 H_1} R(s)$$

This is a simplified version of the formula given in the *NCEES Handbook*, Instrumentation, Measurement, and Control section. Apply the formula to this problem. Substituting the values from the problem statement, the transfer function of the system is

$$\frac{Y(s)}{R(s)} = \frac{s+5}{1+(s+5)^2}$$

The Laplace transform pairs can be used to solve nonhomogeneous differential equations by means of simple algebra. From the Laplace Transform Pairs table given in the *NCEES Handbook*, the applicable pair is

$$\text{if } f(t) = e^{-\alpha t}\cos \beta t$$

$$\text{then } F(s) = \frac{(s+\alpha)}{(s+\alpha)^2 + \beta^2}$$

In this case,

$$\begin{aligned}\alpha &= 5 \\ \beta &= 1\end{aligned}$$

Using the inverse Laplace transform,

$$\begin{aligned}f(t) &= L^{-1}\left(\frac{s+5}{(s+5)^2+1}\right) \\ &= e^{-5t}\cos t\end{aligned}$$

The answer is (C).

101.

step 1: The power is given in kilowatts. Use the *NCEES Handbook*, Units and Conversion Factors section.

$$\begin{aligned}1 \text{ W} &= 1 \text{ J} \\ &= 1 \text{ N·m} \\ 150 \text{ kW} &= 150 \text{ kN·m}\end{aligned}$$

step 2: See the *NCEES Handbook*, Dynamics of Mechanisms section. Analyze the force system using the gear force diagram as shown.

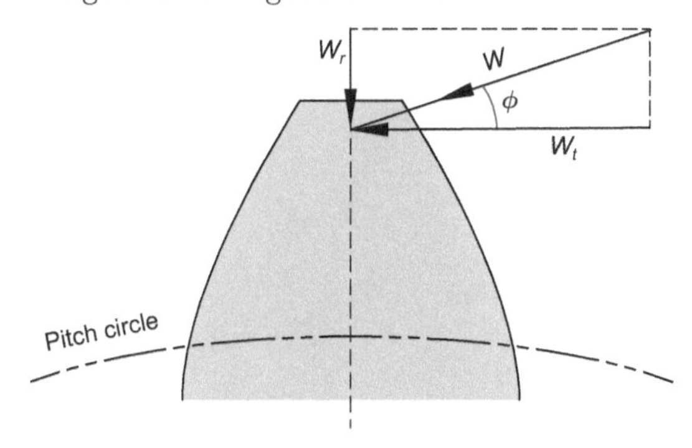

The load, W, on a gear has two components: tangential and radial, as shown. The tangential component, W_t, transmits torque from one gear to the other. The load-power relationships are

$$\begin{aligned}W_t &= W\cos 20° \\ T &= \text{Force} \times \text{lever arm} \\ &= W_t \times \text{radius}\end{aligned}$$

$$\begin{aligned}P &= T\omega \\ &= 150 \text{ kN·m} \\ T &= \frac{150 \text{ kN·m}}{\omega}\end{aligned}$$

step 3: Determine the angular velocity.

$$\begin{aligned}\omega &= \left(2\pi\ \frac{\text{rad}}{\text{rev}}\right)\left(600\ \frac{\text{rev}}{\text{min}}\right)\left(\frac{1\ \text{min}}{60\ \text{s}}\right)\\ &= 20\pi\ \text{rad/s}\end{aligned}$$

step 4: Determine the tangential load.

$$\begin{aligned}W_t &= \frac{2T}{d}\\ &= \left(\frac{(2)(2387\ \text{N·m})}{300\ \text{mm}}\right)\left(\frac{1000\ \text{mm}}{1\ \text{m}}\right)\left(\frac{1\ \text{kN}}{1000\ \text{N}}\right)\\ &= 15.913\ \text{kN}\end{aligned}$$

step 5: Determine the total load.

$$\begin{aligned}W &= \frac{W_t}{\cos\phi}\\ &= \frac{15.913\ \text{kN}}{\cos 20^\circ}\\ &= 16.93\ \text{kN}\quad (17\ \text{kN})\end{aligned}$$

The answer is (B).

102. Use the *NCEES Handbook*, Variable Loading Failure Theories section. The fatigue failure envelope using the modified Goodman theory is

$$\frac{\sigma_a}{S_e} + \frac{\sigma_m}{S_{ul}} \geq 1$$

It is given that

$$\begin{aligned}S_{ut} &= 1200\ \text{MPa}\\ \sigma_a &= 60\ \text{MPa}\\ \sigma_m &= 60\ \text{MPa}\end{aligned}$$

Use the data above. For the failure envelope,

$$\frac{\sigma_a}{S_e} + \frac{\sigma_m}{S_{ut}} \geq 1$$

$$\frac{60\ \text{MPa}}{S_e} + \frac{60\ \text{MPa}}{1200\ \text{MPa}} = 1$$

$$S_e = 63.2\ \text{MPa}\quad (63\ \text{MPa})$$

The answer is (A).

103. See the *NCEES Handbook*, Mechanical Design and Analysis: Springs section for equivalent spring constant. Stiffness is defined as the load required to cause unit deformation.

step 1: The figure shows that two elements undergo equal displacement and are connected in parallel to carry the load. Both elements share in resisting the point load. Therefore,

$$P = P_1 + P_2$$

The portion of load each element resists is proportional to its stiffness.

step 2: Determine the stiffness, k_1, of element 1. Element 1 is a simply supported beam acting as a leaf-type spring. See the Mechanics of Materials section for deflection/stiffness.

$$\begin{aligned}\Delta_1 &= \frac{P_1 L^3}{48EI}\\ &= \frac{P_1\left((4\ \text{ft})\left(\frac{12\ \text{in}}{1\ \text{ft}}\right)\right)^3}{(48)\left(29{,}000\ \frac{\text{kips}}{\text{in}^2}\right)(0.21\ \text{in}^4)}\\ &= 0.378P_1\ \text{in}\end{aligned}$$

step 3: Element 2 is a linear spring, and its displacement under P_2 is

$$\begin{aligned}\Delta_2 &= \frac{P_2}{k_2} = \frac{P_2}{4\ \frac{\text{kips}}{\text{in}}}\\ &= 0.25P_2\ \text{in}\end{aligned}$$

step 4: Equate the two displacements and determine P_2.

$$\begin{aligned}0.378P_1 &= 0.25P_2\\ P_1 &= \frac{0.25P_2}{0.378} = 0.66P_2\end{aligned}$$

From step 1, it is known that $P_1 + P_2 = 1$ kip.

$$\begin{aligned}0.66P_2 + P_2 &= 1.66P_2 = 1\ \text{kip}\\ P_2 &= \frac{1\ \text{kip}}{1.66}\\ &= 0.6024\ \text{kips}\quad (60\%)\end{aligned}$$

The answer is (B).

104. See the *NCEES Handbook*, Mechanical Engineering section.

step 1: The shear stress, τ, in the helical linear spring is

$$\tau = K_s \frac{8FD}{\pi d^3}$$

Determine the parameters needed on the right side of equation, keeping consistent units.

$$\begin{aligned} \text{Applied force, } F &= 1 \text{ kip} \\ \text{Spring mean dia., } D &= 12 \text{ in} \\ \text{Spring wire dia., } d &= 1 \text{ in} \\ \text{Diameter ratio, } C &= D/d = 12 \text{ in}/1 \text{ in} = 12 \\ K_s &= \frac{2C+1}{2C} = \frac{2 \times 12 + 1}{2 \times 12} = \frac{25}{24} \end{aligned}$$

step 2: Plug the above parameters into the shear stress formula.

$$\begin{aligned} \tau &= K_s \frac{8FD}{\pi d^3} \\ &= \left(\frac{25}{24}\right)\left(\frac{8(1 \text{ kip})(12 \text{ in})}{\pi (1 \text{ in})^3}\right) \\ &= 31.83 \text{ kips/in}^2 \quad (32 \text{ ksi}) \end{aligned}$$

The answer is (D).

105. A cylindrical vessel can be classified as a thin-walled or thick-walled vessel. If the thickness of the cylinder wall is about 1/10 or less of the inside radius, the cylinder is considered thin-walled. In this problem, it is given that the cylindrical vessel is thin-walled. A cylindrical pressure vessel under an internal pressure is subjected to two types of stresses: axial and radial (or hoop) stresses. For a thin-walled cylinder under internal pressure, its hoop stress is double the axial stress. Therefore, in designing the wall thickness of a cylinder, tangential (hoop) stress controls.

step 1: See the *NCEES Handbook*, Cylindrical Pressure Vessel section and apply the formula

$$\sigma_t = \frac{P_i r}{t}$$

step 2: Keep the units consistent.

$$\begin{aligned} P_i &= 125 \text{ psi} \\ \sigma_t &= 24 \text{ ksi} = 24{,}000 \text{ psi} \\ r &= \frac{48 \text{ in}}{2} = 24 \text{ in} \end{aligned}$$

step 3: Apply the known values to determine the thickness.

$$\begin{aligned} \sigma_{t,\max} &= \frac{P_i r}{t_{\min}} \\ 24{,}000 \text{ psi} &= \frac{(125 \text{ psi})(24 \text{ in})}{t_{\min}} \\ t_{\min} &= \frac{(125 \text{ psi})(24 \text{ in})}{24{,}000 \text{ psi}} \\ &= 0.125 \text{ in} \quad (1/8 \text{ in}) \end{aligned}$$

step 4: Check if the calculated wall thickness complies with the thin-walled vessel assumption. The maximum wall thickness for thin-walled cylinder is

$$\begin{aligned} t_{\max} &= \frac{r_{\text{inside}}}{10} \\ r_{\text{inside}} &= \frac{\bar{d} - t_{\min}}{2} \\ &= \frac{48 \text{ in} - 0.125 \text{ in}}{2} \\ &= 23.937 \text{ in} \end{aligned}$$

Determine the permissible wall thickness.

$$\begin{aligned} t_{\max} &= \frac{23.937 \text{ in}}{10} \\ &= 2.39 \text{ in} > \frac{1}{8} \text{ in} \end{aligned}$$

The permissible maximum wall thickness using the thin-wall design method is more than the calculated thickness. Therefore, the design is valid.

The answer is (A).

106. Use the *NCEES Handbook*, Bearings section. The minimum basic load rating for ball bearing, C, is given by

$$\begin{aligned} C &= PL^{1/3} \\ P &= \text{design radial load} = 3.9 \text{ kN} \\ L &= \text{design life (mil. rev.)} \end{aligned}$$

step 1: Determine design life, L, in terms of revolutions.

$$\begin{aligned} N &= 600 \text{ rpm} \\ \text{Rating life} &= 30{,}000 \text{ hr} \end{aligned}$$

$$\begin{aligned} L &= \frac{60NL}{10^6} \\ &= \frac{\left(60\ \frac{\text{min}}{\text{hr}}\right)\left(600\ \frac{\text{rev}}{\text{min}}\right)(30{,}000\ \text{hr})}{10^6} \\ &= 1080\ \text{mil. rev.} \end{aligned}$$

step 2: Determine load rating, C.

$$\begin{aligned} C &= PL^{1/3} = (3900\ \text{N})(1080\ \text{mil. rev.})^{1/3} \\ &= 40{,}013 \quad (40{,}000) \end{aligned}$$

The answer is (C).

107. Use the *NCEES Handbook*, Screw Thread section. The torque, M, required in raising a load P using a screw-jack with square thread is

$$M = Pr\tan(\alpha + \phi)$$

From the Mathematics section of the *NCEES Handbook*, use the identity

$$\tan(\alpha + \phi) = \frac{\tan\alpha + \tan\phi}{1 - \tan\alpha\tan\phi}$$

The coefficient of friction is $\mu = \tan\ \phi = 0.12$, and α is the pitch angle.

$$\begin{aligned} \tan\alpha &= \frac{p}{2\pi r} \\ &= \frac{8\ \text{mm}}{2\pi(30\ \text{mm})} \\ &= 0.042 \end{aligned}$$

$$\begin{aligned} M &= Pr\frac{\tan\alpha + \tan\phi}{1 - \tan\alpha\tan\phi} \\ M &= (500\ \text{kN})(0.03\ \text{m})\left(\frac{0.042 + 0.12}{1 - (0.042)(0.12)}\right) \\ &= (15\ \text{kN·m})\left(\frac{0.162}{0.995}\right) \\ &= 2.442\ \text{kN·m} \quad (2.4\ \text{kN·m}) \end{aligned}$$

The answer is (B).

108. The force on the drive side of the belt is given as 200 lbf. The other side of the belt is also called the slack side. See the *NCEES Handbook*, Belt Friction section.

step 1: Apply the belt force transmission formula.

$$F_1 = F_2 e^{\mu\theta}$$

F_1 is 200 lbf force being applied in the direction of impending motion (drive side or tight side), and F_2 is the force applied to resist impending motion (slack side).

The coefficient of friction and the contact angle between belt and pulley (in radians) are given as

$$\mu = 0.24$$

$$\begin{aligned} \theta &= 180° \\ &= \pi\ \text{rad} \end{aligned}$$

step 2: Using the above parameters, solve the equation to determine the unknown tension force F_2.

$$\begin{aligned} F_1 &= F_2 e^{0.24\pi} \\ &= 2.125 F_2 \\ F_2 &= \frac{200\ \text{lbf}}{2.125} \\ &= 94.12\ \text{lbf} \quad (94\ \text{lbf}) \end{aligned}$$

The answer is (B).

109. Three modes of failure of a riveted joint are shown in the *NCEES Handbook*, Mechanical Engineering: Joining Methods section.

In this case, it is given that the joint rivets are oversized. Therefore, the rivets will not fail first; option A is incorrect. Both plates have equal thickness. Plate B is wider than plate A. Therefore, plate A is weaker and will fail before plate B does. Options C and D do not apply. Therefore, plate A will rupture or tear off, and the assembly will fail. Option B is correct.

The answer is (B).

110. The terms basic size and nominal size are used interchangeably. In any production process, no component can be manufactured precisely to a given dimension; it can only be made to lie between two limits, upper and lower. Both deviations are defined in the *NCEES Handbook*. In this case, the deviations for the shaft are given. According to the metric standard, lowercase letters are used when referring to the shaft. Therefore,

$$d = 20 \text{ mm}$$
$$\delta u = 0.03 \text{ mm}$$
$$\delta l = 0.04 \text{ mm}$$

The *NCEES Handbook* defines fundamental deviation, δ_F, as the upper or lower deviation depending on which is closer to the basic size. In other words, it is the **minimum difference in size between a component and the basic size**. This is identical to the upper deviation for shafts and the lower deviation for holes. Therefore,

$$\delta_F = 0.03 \text{ mm}$$

$$\begin{aligned} d_{\text{max}} &= d + \delta_F \\ &= 20 \text{ mm} + 0.03 \text{ mm} \\ &= 20.03 \text{ mm} \end{aligned}$$

The answer is (B).

Solutions
Exam 2

111. Geothermal energy is generated when water comes into contact with heated underground rocks. The heat turns the water into a steam which can be used to generate power. In this problem, the thermal gradient, m, is constant because the thermal variation along the depth is linear. Calculate the slope, m.

$$\begin{aligned} m &= \frac{y_2 - y_1}{x_2 - x_1} = \frac{\text{temperature difference}}{\text{depth difference}} \\ &= \frac{90°\text{C} - 20°\text{C}}{2000 \text{ m} - 0 \text{ m}} \\ &= 0.035 \text{ °C/m} \end{aligned}$$

Use the point-slope form of the straight-line equation.

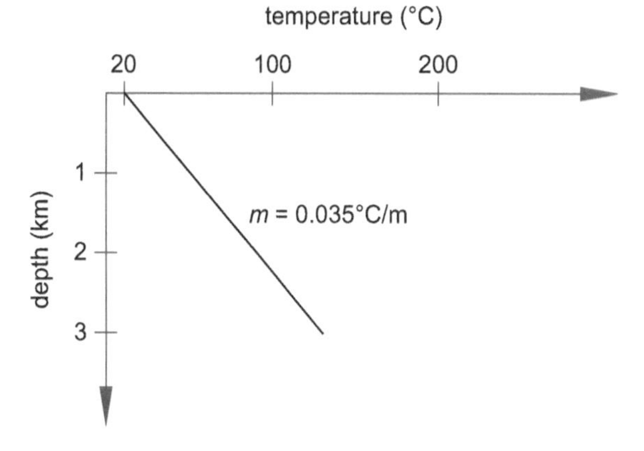

$$\begin{aligned} y - y_1 &= m(x - x_1) \\ T_{3800} - T_0 &= m(3800 \text{ m} - 0 \text{ m}) \\ T_{3800} - 20°\text{C} &= (0.035 \text{ °C/m})(3800 \text{ m} - 0 \text{ m}) \end{aligned}$$

Therefore, the temperature at a depth of 3800 m can be determined as

$$\begin{aligned} T_{3800} &= (0.035 \text{ °C/m})(3800 \text{ m}) + 20°\text{C} \\ &= 153°\text{C} \end{aligned}$$

The answer is (C).

112. The velocity, v, is defined as the distance travelled per unit time.

$$\text{v} = \frac{ds}{dt} = \frac{d}{dt}\left(\frac{t^2 + 1}{t + 1}\right)$$

Use the *NCEES Handbook*, Mathematics: Derivatives section. Use the quotient rule.

$$\frac{d\left(\frac{u}{\text{v}}\right)}{dx} = \frac{\text{v}\left(\frac{du}{dx}\right) - u\left(\frac{d\text{v}}{dx}\right)}{\text{v}^2}$$

Let $u = t^2 + 1$ and $\text{v} = t + 1$.

$$\begin{aligned} \frac{du}{dt} &= \frac{d(t^2 + 1)}{dt} = 2t \\ \frac{d\text{v}}{dt} &= \frac{d(t + 1)}{dt} = 1 \end{aligned}$$

Substitute these values into the velocity equation.

$$\begin{aligned} \text{v} &= \frac{(t + 1)(2t) - (t^2 + 1)(1)}{(t + 1)^2} \\ &= \frac{(t^2 + 2t - 1)}{(t + 1)^2} \end{aligned}$$

For $t = 2.5$ s, find the velocity.

$$\begin{aligned} \text{v} &= \frac{\left((2.5 \text{ s})^2 + 2(2.5 \text{ s}) - 1\right)}{(2.5 \text{ s} + 1)^2} \\ &= \frac{10.25}{12.25} \\ &= 0.8367 \quad (0.84) \end{aligned}$$

The answer is (B).

113. The equation is a second-order linear homogeneous differential equation with constant real coefficients.

For the form $y'' + ay' + b = 0$, the general solution is

$$y = Ce^{rx}$$

Substituting this solution in the equation gives

$$\begin{aligned} r^2 + ar + b &= 0 \\ r^2 + 3r + 2 &= 0 \\ (r + 2)(r + 1) &= 0 \end{aligned}$$

The roots are $r_1 = -2$ and $r_2 = -1$.

Since the roots are real, the solution is of the form

$$y = C_1 e^{r_1 x} + C_2 e^{r_2 x}$$

Substituting the values of r_1 and r_2, the solution is

$$y = C_1 e^{-2x} + C_2 e^{-x}$$

The answer is (A).

114. To multiply two matrices, the number of columns in the first matrix should equal the number of rows in the second matrix. In this case, the condition is satisfied because matrix A has two columns and matrix B has two rows. Therefore, the multiplication of matrices A and B is possible. Option D is incorrect. The product matrix will have as many rows as the first matrix and as many columns as the second matrix. Therefore, the product matrix C will have three rows and three columns.

The answer is (C).

115. See the *NCEES Handbook*, Mathematics section, for Newton's method of root extraction. The equation can be simplified as

$$\begin{aligned} a^{j+1} &= a^j - \left.\frac{f(x)}{\frac{df(x)}{dx}}\right|_{x=a^j} \\ &= a^j - \left.\frac{f(x)}{f'(x)}\right|_{x=a^j} \end{aligned}$$

The initial estimated root value, $a_0 = 2$, is near enough to the actual root that the algorithm would converge.

step 1: Find the derivative of the function.

$$\begin{aligned} f(x) &= x^7 - 100 \\ f'(x) &= 7x^6 \end{aligned}$$

The zero root is given as

$$a_0 = 2$$

step 2: Use the zeroth root value and determine the first root.

$$\begin{aligned} a_1 &= 2 - \frac{(2)^7 - 100}{7(2)^6} \\ &= 1.9375 \end{aligned}$$

step 3: Use the first root value and determine the second root.

$$\begin{aligned} a_2 &= 1.9375 - \frac{(1.9375)^7 - 100}{7(1.9375)^6} \\ &= 1.930769 \quad (1.9308) \end{aligned}$$

The answer is (C).

116. The algorithm has an endless loop. It keeps on iterating or looping; it never reaches the END statement. The missing print statement is not considered an error.

The answer is (B).

117. Use the *NCEES Handbook*, Mathematics: Partial Derivative section. Use the differential chain rule to determine the differential.

$$\begin{aligned} \frac{\partial z}{\partial s} &= \frac{\partial z}{\partial x}\frac{\partial x}{\partial s} + \frac{\partial z}{\partial y}\frac{\partial y}{\partial s} \\ &= y(2s) + x\left(\frac{1}{t}\right) \\ &= \left(\frac{s}{t}\right)(2s) + (s^2 + t^2)\left(\frac{1}{t}\right) \\ &= \frac{3s^2 + t^2}{t} \end{aligned}$$

The answer is (A).

118. Calculate the value of the integral.

$$\begin{aligned} \int_{-2}^{1} (3x^2 + 2x - 9)\,dx &= x^3 + x^2 - 9x\Big|_{-2}^{1} \\ &= \left(1^3 + 1^2 - 9(1)\right) - \left((-2)^3 \right. \\ &\quad \left. + (-2)^2 - 9(-2)\right) \\ &= -21 \end{aligned}$$

The answer is (A).

119. The order in which workers are combined in a crew is not considered in group formation. The combination of n objects taken r at a time is expressed as

$$C(n,r) = \frac{n!}{r!\,(n-r)!}$$

In this case one man is selected from 8 men, and 4 women are selected from 12 women. Thus, the number of combinations is

$$\begin{aligned} N &= C(8,1)C(12,4) \\ &= \left(\frac{8!}{1!\,(8-1)!}\right)\left(\frac{12!}{4!\,(12-4)!}\right) \\ &= \left(\frac{8!}{1!\,(7)!}\right)\left(\frac{12!}{4!\,(8)!}\right) \\ &= 3960 \end{aligned}$$

The answer is (A).

120. Calculate the mean of the strengths.

$$\begin{aligned} \overline{X} &= \left(\frac{4450\text{ psi} + 4675\text{ psi} + 4898\text{ psi} + 4120\text{ psi}}{4}\right) \\ &= 4536\text{ psi} \end{aligned}$$

The sample variance, s^2, is calculated as

$$\begin{aligned} s^2 &= \frac{1}{n-1}\left(\sum_{i=1}^{n}(X_i - \overline{X})^2\right) \\ &= \frac{1}{4-1}\left(\begin{array}{l} (4450\text{ psi} - 4536\text{ psi})^2 \\ \quad +(4675\text{ psi} - 4536\text{ psi})^2 \\ \quad +(4898\text{ psi} - 4536\text{ psi})^2 \\ \quad +(4120\text{ psi} - 4536\text{ psi})^2 \end{array}\right) \\ &= 110{,}272\text{ psi}^2 \end{aligned}$$

Determine the simple standard deviation.

$$\begin{aligned} s &= \sqrt{110{,}272\text{ psi}^2} \\ &= 332\text{ psi} \quad (330\text{ psi}) \end{aligned}$$

The answer is (B).

121. Calculate the expected cost of each estimate.

$$\text{Cost}_{\text{total}} = A_{\text{sandy silt}}\text{Cost}_{\text{sandy silt}} + A_{\text{rock}}\text{Cost}_{\text{rock}}$$

$$\begin{aligned} \text{Cost}_A &= \left((25\%)(500{,}000\text{ ft}^2)\right)\left(100\ \frac{\$}{\text{ft}^2}\right) \\ &\quad +\left((100\% - 25\%)(500{,}000\text{ ft}^2)\right)\left(300\ \frac{\$}{\text{ft}^2}\right) \\ &= \$125\text{M} \end{aligned}$$

$$\begin{aligned} \text{Cost}_B &= \left((55\%)(500{,}000\text{ ft}^2)\right)\left(100\ \frac{\$}{\text{ft}^2}\right) \\ &\quad +\left((100\% - 55\%)(500{,}000\text{ ft}^2)\right)\left(300\ \frac{\$}{\text{ft}^2}\right) \\ &= \$95\text{M} \end{aligned}$$

The weighted arithmetic mean of a set of values can be determined using the expression from the Engineering Probability and Statistics section of the *NCEES Handbook.*

$$\overline{X}_w = \frac{\sum w_i X_i}{\sum w_i}$$

In this case, since one report is given twice the weight, $w_B = 1$, and $w_A = 2$. The weighted expected cost is

$$\begin{aligned} \overline{X}_w &= \frac{w_A\text{Cost}_A + w_B\text{Cost}_B}{w_A + w_B} \\ &= \frac{2(\$125\text{M}) + 1(\$95\text{M})}{(2+1)} \\ &= \$115\text{M} \end{aligned}$$

The answer is (D).

122. Use the formulas given in the Linear Regression and Goodness of Fit section of the *NCEES Handbook.* To evaluate the parameters in the regression equations, tabulate the data and use the summed values in the formulas.

$$\begin{aligned} S_{\text{xy}} &= \sum_{i=1}^{n} x_i y_i - \left(\frac{1}{n}\right)\left(\sum x_i\right)\left(\sum y_i\right) \\ &= 324 - \left(\frac{1}{4}\right)(19)(57) \\ &= 53.25 \end{aligned}$$

$$\begin{aligned} S_{\text{xx}} &= \sum_{i=1}^{n}(x_i)^2 - \left(\frac{1}{n}\right)\left(\sum x_i\right)^2 \\ &= 119 - \left(\frac{1}{4}\right)(19)^2 \\ &= 28.75 \end{aligned}$$

$$\sum_{i=1}^{4} y_i^2 = 9^2 + 11^2 + 15^2 + 22^2 = 911$$

$$S_{yy} = \sum_{i=1}^{n} y_i^2 - \left(\frac{1}{n}\right)\left(\sum_{i=1}^{n} y_i\right)^2$$
$$= 911 - \left(\frac{1}{4}\right)(57)^2$$
$$= 98.75$$

$$\text{MSE} = \frac{S_{xx}S_{yy} - S_{xy}^2}{S_{xx}(n-2)}$$

$$S_e^2 = \frac{(28.75)(98.75) - (53.25)^2}{(28.75)(4-2)}$$
$$= 0.06087 \quad (0.061)$$

The answer is (A).

123. As noted in the probability and density functions table in the Engineering Probability and Statistics section of the *NCEES Handbook*, a function is said to be normally distributed if its density function is given by an expression of the form

$$f(x) = \frac{1}{\sigma\sqrt{2\pi}} e^{-\frac{1}{2}\left(\frac{x-\mu}{\sigma}\right)^2}$$

The given function is in the form of the normal distribution, with $\mu = 15$ and $\sigma = 1$. Integrate the function to obtain the fraction

$$X = \int_{14}^{16} \frac{1}{\sqrt{2\pi}} e^{-\frac{(x-15)^2}{2}}$$

The nondimensional parameter used in the unit normal distribution tables given in the Engineering Probability and Statistics section of the *NCEES Handbook* is

$$z = \frac{x-\mu}{\sigma}$$

For $x = 14$,

$$z = (14 - 15)/1 = -1$$

For $x = 16$,

$$z = (16 - 15)/1 = 1$$

Use the unit normal distribution method with tables. The unit normal distribution curve with the parameter z along its x-axis is shown. The area between $z = -1$ and $z = 1$ represents the area from 14 in to 16 in of rainfall.

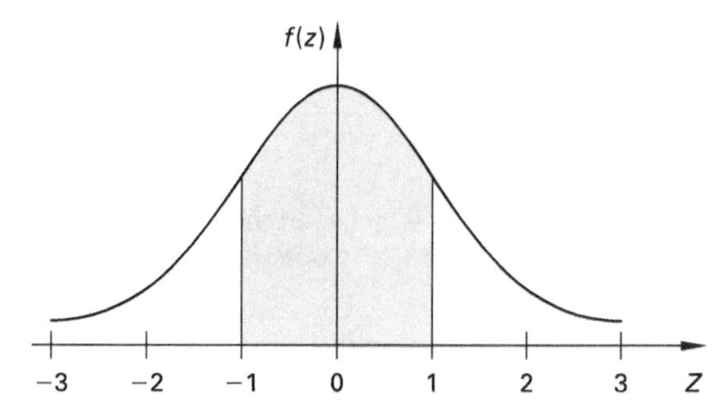

From the Unit Normal Distribution table in the *NCEES Handbook*, the shaded area is

$$F(1 \text{ and } -1) = W(1) = 0.6827 \quad (0.68)$$

The answer is (C).

124. Ethics are a set of guidelines, rules, philosophical concepts, customs, norms, and aspirations for a licensee to follow. The Code articulates the ways in which moral and ethical principles apply to unique situations encountered in professional practice. It indicates to others that the professionals are seriously concerned about responsible and professional conduct. In some cases, it is impossible to comply with every aspect of the Code. Therefore, ethics are also called a set of aspirations that a licensed engineer should aim for. However, ethics are not subject to the law. As the late Chief Justice of the U.S. Supreme Court Earl Warren put it, "Society would come to grief without ethics, which is unenforceable in the courts and cannot be made part of law. . . Not only does law in a civilized society presuppose ethical commitment, it presupposes the existence of a broad area of human conduct controlled only by ethical norms and not subject to law at all."

The answer is (A), (B), and (C).

125. A patent for an invention is the grant of a property right to the inventor, issued by the USPTO. According to the USPTO, there are three types of patents.

1. Utility patents may be granted to anyone who invents or discovers any new and useful process, machine, article of manufacture, or composition of matter, or any new and useful improvement thereof;

2. Design patents may be granted to anyone who invents a new, original, and ornamental design for an article of manufacture; and

3. Plant patents may be granted to anyone who invents or discovers and asexually reproduces any distinct and new variety of plant.

Other than these three, there is no other patent category.

The answer is (A), (B), and (C).

126. See the *NCEES Handbook*, Ethics and Professional Practice: Intellectual Property section. One of the requirements to be protected under copyright laws is that the work must be tangible. Ideas are intangible and cannot be copyrighted. Therefore, statement I is incorrect.

To be protected under copyright laws, the work must be original, whether published or unpublished. Therefore, statement II is correct.

To be protected under copyright laws, the work must be original, whether published or unpublished. In this case, the engineer has her intellectual property right over the improvements she designed and implemented in the new machine. Her original work is not considered as copyright infringement. Therefore, statement III is correct.

As stated in the *NCEES Handbook*, a utility patent may be granted to anyone who discovers any new machine or article of manufacture. As such, she may be entitled to a utility patent to the extent of her discovery and invention. Statement IV is correct.

The answer is (A).

127. See the *NCEES Handbook*, Ethics and Professional Practice *Model Rules*, Section 240.15 Rules of Professional Conduct. Engineers are required to perform their professional services at a reasonable standard of care. If an engineer falls short of this standard, the engineer can be found negligent, lacking competence, and committing errors and omission.

An engineer's liability of can be divided into two categories: professional liability and general liability. Accordingly, different insurance policies are needed to cover a claim arising out of acts and omission of the insured. As a buffer against the professional liability, an engineering firm can purchase insurance to protect itself and its employees. It is called errors and omissions (E&O) insurance, professional liability (PL) insurance, negligence insurance, or malpractice insurance. With the payment of the required premium, the E&O insurance protects the firm and its design professional employees from the full cost of defending against a negligence claim made by a client or others. Option D is correct. The insurance types listed in options A through C cover general or nonprofessional situations. For example, a firm can procure auto insurance to cover liabilities resulting from the employees' use of their vehicles to conduct activities within the scope of their employment.

The answer is (D).

128. Meeting the minimum requirements of a building code may not yield sound design documents. The documents must conform to the accepted engineering standards in order to safeguard the life, health, property, and welfare of the public. Therefore, option A is incorrect.

According to the *NCEES Handbook*, Ethics and Professional Practice: *Model Rules*, Section 240.15 Rules of Professional Conduct, A, it is the paramount duty of an engineer to safeguard the life, health, and welfare of the public in conducting their professional service. Therefore, B is a correct option.

According to the *NCEES Handbook*, Ethics and Professional Practice: *Model Rules*, Section 240.15 Rules of Professional Conduct, B.2, licensees should meet two requirements before affixing their seal: (1) plans or documents must be prepared under their responsible charge and (2) must have competence in the area of design. Therefore, option C is incorrect.

Unless required by the contract with their client, there is no ethical requirement for a licensee to specify only green or environmentally friendly materials in design. According to the *NCEES Handbook*, Ethics and Professional Practice: Societal Considerations section, the engineers must consider sustainable principles, as a part of the societal considerations. Therefore, option D is incorrect.

The answer is (B).

129. A licensee who is an employee should inform the licensee's employer. A self-employed licensee should inform the licensee's client. In addition, a licensee should inform other authorities as appropriate. See *Model Rules*, Section 240.15 Rules of Professional Conduct, A.3, in the *NCEES Handbook*.

The answer is (D).

130. The firm would earn a uniform amount, A, of \$100,000/yr from now to perpetuity. The amount, A, is assumed to be inflation-adjusted. It would not increase since there is no growth. The rate of return, i, is 10%, or 0.1. The Engineering Economics section of the *NCEES Handbook* provides the capitalized costs formula. Using the formula, the present value of the firm is

$$\begin{aligned} P &= \frac{A}{i} \\ &= \frac{\$100{,}000}{0.1} \\ &= \$1{,}000{,}000 \end{aligned}$$

The answer is (B).

131. The initial cost and annual cost are both expenses. The rate of return (ROR) on the investment is unknown. No factor table is available to determine ROR precisely. However, it can be determined by interpolating between two applicable factor tables. Equate the expenses and returns (in thousand dollars).

$$40+3\left(\frac{P}{A},i,5\right)=15\left(\frac{P}{A},i,5\right)$$
$$\left(\frac{P}{A},i,5\right)=\frac{40}{12}=3.3333$$

The factor tables for 12% and 18% provide P/A factors from which the required ROR can be interpolated.

$$\left(\frac{P}{A},12\%,5\right)=3.6048$$
$$\left(\frac{P}{A},18\%,5\right)=3.1272$$

Use linear interpolation between the 12% and 18% interest rates and their ROR. Use the line equation from the Mathematics section of the *NCEES Handbook*.

$$m=\frac{y_2-y_1}{x_2-x_1}=\frac{3.1272-3.6048}{0.18-0.12}=-7.96$$
$$\begin{aligned}x_{\text{actual}}&=\frac{y_{\text{actual}}-y_1}{m}+x_1\\&=\frac{3.3333-3.6048}{-7.96}+0.12\\&=0.1541\quad(15\%)\end{aligned}$$

The answer is (C).

132. Assume that the concrete mix contains x units of admixture A and y units of admixture B. Calculate the mix cost, W, as

$$\begin{aligned}W&=xA+yB\\&=\left(\frac{\$4}{\text{dose}}\right)(\text{doses of }A)+\left(\frac{\$12}{\text{dose}}\right)(\text{doses of }B)\end{aligned}$$

The mix requires both admixtures. To keep the mix cost to a minimum, three conditions must be satisfied.

$$3\le A\le 9$$
$$B\ge 8$$
$$A+B\le 16$$

Graph the conditions as shown. The shaded area bound by the three points marked 1, 2, and 3 represents the limits that satisfy the inequalities. The mix cost W lies within or at the boundaries of the shaded area. The value of W closest to the origin would yield the minimum mix cost. Compute the cost for the three corners of the shaded area. By inspection, the corner (3,8) is located closest to the origin and provides the least expensive mix design.

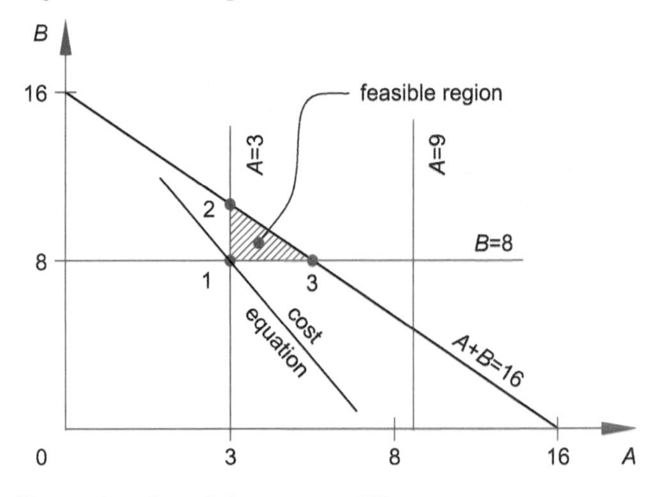

Determine the minimum cost, W_{min}, as

$$\begin{aligned}W_{\text{min}}&=(\$4)(3)+(\$12)(8)\\&=\$108\end{aligned}$$

The answer is (B).

133. The cost grows linearly, as shown.

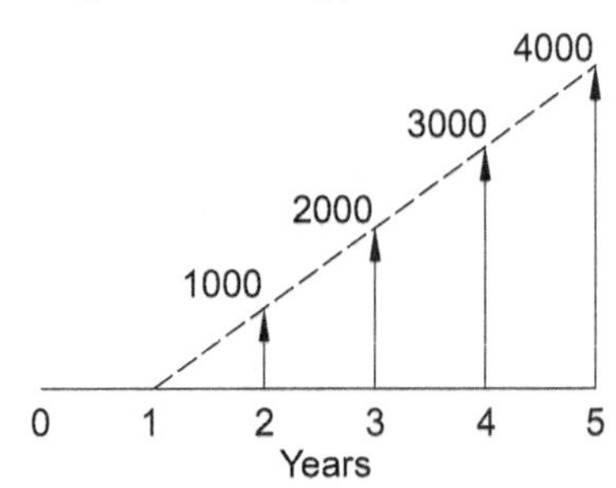

The growth increment is $G=\$1000$. From the factor table,

$$\left(\frac{A}{G},8\%,5\right)=1.8465$$
$$\begin{aligned}A&=(\$1000)(1.8465)\\&=\$1846.5\quad(\$1800)\end{aligned}$$

The answer is (A).

134. The monthly return is 0.5%. The number of monthly deposits in 5 years will be 60. Use the factor tables to determine the monthly deposit as

$$\begin{aligned}A&=\$500{,}000\left(\frac{A}{F},\,0.5\%,\,60\right)\\&=\$500{,}000(0.0143)\\&=\$7150\end{aligned}$$

The answer is (A).

135. For the project to be feasible, its present worth must be positive (greater than zero). Let the annual income from the project be X, as shown.

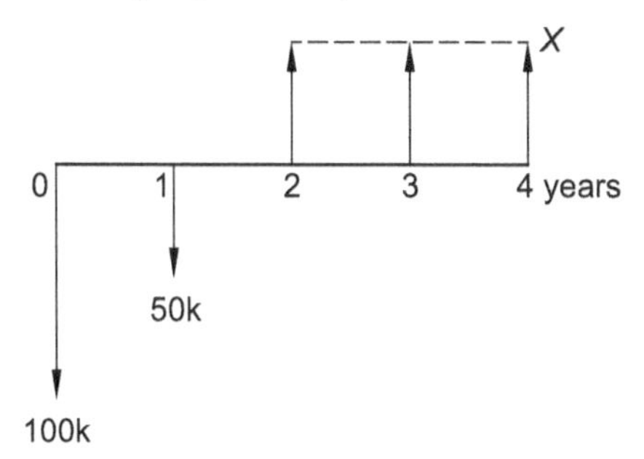

For a 10% annual return, use factor tables to determine present worth as

$$\begin{aligned} P &= -\$100{,}000 - \$50{,}000\left(\frac{P}{F}, 10\%, 1\right) \\ &\quad + X\left(\left(\frac{P}{F}, 10\%, 2\right) + \left(\frac{P}{F}, 10\%, 3\right)\right. \\ &\quad \left. + \left(\frac{P}{F}, 10\%, 4\right)\right) \\ &= -\$100{,}000 - \$50{,}000(0.9091) + X(0.8264) \\ &\quad + X(0.7513) + X(0.6830) \\ &= -100{,}000 - 45{,}455 + 2.2607X \\ 0 &< -100{,}000 - 45{,}455 + 2.2607X \\ X &> \$64{,}341 \quad (\$64{,}000) \end{aligned}$$

The answer is (A).

136. The following information is given.

$$\begin{aligned} N &= 400 \text{ turns} \\ \mu &= 8.8 \times 10^{-3} \text{ H/m} \\ A &= 2 \text{ cm} \times 2 \text{ cm} = 4 \text{ cm}^2 = 0.0004 \text{ m}^2 \\ l &= \pi D = \pi(0.3 \text{ m}) \end{aligned}$$

See the *NCEES Handbook,* Electrical and Computer Engineering section. The inductance L of a coil of N turns wound on a core with cross-sectional area A and permeability μ with a mean path of l is calculated as

$$\begin{aligned} L &= \frac{N^2\mu A}{l} \\ &= \frac{(400)^2(8.8 \times 10^{-3} \text{ H/m})(0.0004 \text{ m}^2)}{0.3\pi \text{ m}} \\ &= 0.5976 \text{ H} \quad (0.6 \text{ H}) \end{aligned}$$

The answer is (B).

137. See the *NCEES Handbook*, Electrical and Computer Engineering section. A parallel plate capacitor with an area A, with plates separated a distance d by an insulator with a permittivity ε, has a capacitance of

$$C = \frac{\varepsilon A}{d}$$

The absolute permittivity of air is

$$\varepsilon_0 = 8.854 \times 10^{-12} \text{ F/m}$$

The relatively permittivity of natural rubber is

$$\varepsilon_r = 7$$

Therefore, the permittivity of natural rubber is

$$\begin{aligned} \varepsilon &= \varepsilon_r \times \varepsilon_0 \\ &= 7 \times 8.854 \times 10^{-12} \text{ F/m} \\ &= 61.978 \times 10^{-12} \text{ F/m} \end{aligned}$$

The plate area is calculated as

$$\begin{aligned} A &= (100 \text{ mm})(100 \text{ mm}) \\ &= 10^4 \text{ mm}^2 = 0.01 \text{ m}^2 \end{aligned}$$

Finally, determine the capacitance, C, as

$$\begin{aligned} C &= \frac{\varepsilon A}{d} \\ &= \frac{(61.978 \times 10^{-12} \text{ F/m})(0.01 \text{ m}^2)}{0.011 \text{ m}} \\ &= 56.344 \times 10^{-12} \text{ F} \\ &= 56.344 \ \mu\mu\text{F} \quad (56 \ \mu\mu\text{F}) \end{aligned}$$

The answer is (B).

138. See the *NCEES Handbook* Electrical and Computer Engineering section.

Kirchhoff's current law states that whatever current goes into a node of the circuit must come out of the node. This is stated in statement A.

Kirchhoff's voltage law states that the algebraic sum of the voltages is zero around any loop in the circuit. This is stated in statement B.

The Norton theorem states that any given complicated circuit can be reduced to a single current source with a parallel resistance. Its diagram is given in the *NCEES Handbook.* This is stated in statement C, and it provides one part of the answer.

The Thevenin theorem states that any given complicated circuit can be reduced to a single voltage source in series with a resistance. Its diagram is given in the *NCEES Handbook.* This is stated in statement D, and it provides the second part of the answer.

The answer is

1. Thevenin theorem	D
2. Norton theorem	C

139. The effective or rms value for a periodic waveform with period T is given in the *NCEES Handbook*, Electrical and Computer Engineering section, as

$$V_{\text{rms}} = \left[\frac{1}{T}\int_0^T v^2(t)\,dt\right]^{1/2}$$

step 1: Determine the equation of the voltage at time interval t. The line equation that is passing the origin is

$$y = mx$$

The slope m is

$$\begin{aligned} m &= \frac{y}{x} \\ &= \frac{60\ \text{V}}{2\ \text{sec}} \\ &= 30\ \frac{\text{V}}{\text{sec}} \end{aligned}$$

From the waveform equation,

$$y = 30x$$
$$\text{or } V(t) = 30t$$

step 2: Using the above data, determine the rms value of voltage.

$$\begin{aligned} V_{\text{rms}} &= \left[\frac{1}{T}\int_0^T (v(t))^2\,dt\right]^{1/2} \\ &= \left[\frac{1}{2}\int_0^T (30t)^2\,dt\right]^{1/2} = \left[\frac{1}{2}\int_0^T 900t^2\,dt\right]^{1/2} \\ &= \left[450\int_0^T t^2\,dt\right]^{1/2} = \left[450\left|\frac{t^3}{3}\right|_0^2\right]^{1/2} \\ &= \left[450\left|\frac{2^3}{3} - 0\right|\right]^{1/2} \\ &= \left[450\left|\frac{8}{3} - 0\right|\right]^{1/2} \\ &= [1200]^{1/2} = 34.64\ \text{V} \quad (35\ \text{V}) \end{aligned}$$

The answer is (B).

140. The motor efficiency is defined as the ratio of motor power output to the power input.

$$\begin{aligned} \eta &= \frac{\text{Power Output}}{\text{Power Input}} \\ &= \frac{\text{Power Input} - \text{Losses}}{\text{Power Input}} \end{aligned}$$

See the *NCEES Handbook*, Electrical and Computer Engineering section. Power absorbed by a resistive element is given as

$$P = VI = I^2R$$

The motor circuit is shown.

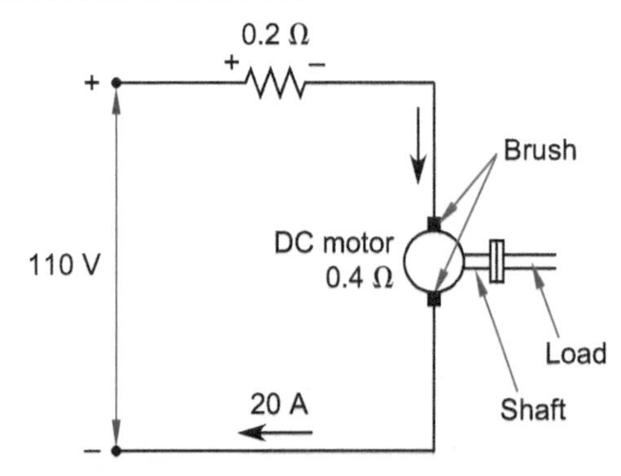

Determine the power input to the motor.

$$\begin{aligned} P_{\text{in}} &= VI = (110\ \text{V})(20\ \text{A}) \\ &= 2200\ \text{W} \end{aligned}$$

The problem states four types of losses.

step 1: Determine power loss in the armature.

$$P_{\text{arm}} = (20\ \text{A})^2(0.4\ \Omega) = 160\ \text{W}$$

step 2: Determine power loss in the field.

$$P_{\text{fld}} = (20\ \text{A})^2(0.2\ \Omega) = 80\ \text{W}$$

step 3: Determine power loss in the brushes.

$$P_{\text{brsh}} = (20\ \text{A})(3\ \text{V}) = 60\ \text{W}$$

It is known that friction loss is 250 W.

step 4: Determine the total power loss.

$$P_{\text{total}} = 160\ \text{W} + 80\ \text{W} + 60\ \text{W} + 250\ \text{W} = 550\ \text{W}$$

step 5: Determine the power output.

$$P_{\text{out}} = 2200\ \text{W} - 550\ \text{W} = 1650\ \text{W}$$

$$\eta = \frac{\text{Power Output}}{\text{Power Input}} = \frac{1650\ \text{W}}{2200\ \text{W}}\left(\frac{100\%}{1}\right) = 75\%$$

The answer is (D).

141. See the *NCEES Handbook*, Statics: Equilibrium Requirements section. For equilibrium to occur under a system of forces, two conditions must be met.

$$\sum F_n = 0$$
$$\sum M_n = 0$$

The problem states the forces are nonparallel. In other words, the forces meet at a point. A system of forces that meet at a point are called concurrent forces. The resultant moment of the system of concurrent forces about their meeting point is zero. Therefore, the forces must be concurrent; otherwise, there would be a resultant moment of the point, and the body would not be in equilibrium.

The answer is (B).

142. A force with its arrow pointing toward a joint denotes a compressive force, and a force with its arrow pointing away from a joint denotes a tensile force. Assume that forces F_2 and F_3 are positive. From equilibrium, the sums of the force components in the x- and y-directions are zero.

$$\sum F_x = F_1\cos\theta - F_4\cos\theta + F_3 = 0$$
$$\sum F_y = F_1\sin\theta - F_4\sin\theta - F_2 = 0$$

The angle θ is determined from the member slope of 3:4.

$$\sin\theta = \frac{3}{5}$$
$$\cos\theta = \frac{4}{5}$$

By substituting the values into the equations,

$$10\ \text{kN}\left(\frac{4}{5}\right) - 20\ \text{kN}\left(\frac{4}{5}\right) + F_3 = 0$$
$$F_3 = 8\ \text{kN}$$

Force F_3 is tensile, as shown. Similarly, from the y-axis,

$$10\ \text{kN}\left(\frac{3}{5}\right) - 20\ \text{kN}\left(\frac{3}{5}\right) - F_2 = 0$$
$$F_2 = -6\ \text{kN}$$

The negative sign shows that F_2 is not tensile but is a compressive force.

The answer is (A).

143. The equivalent loading is the resultant force and accompanying moment. The resultant applied force, $\overline{P}$, is the algebraic sum of all applied forces at point A, as shown. Let upward force be positive.

$$\overline{P} = \sum P = 100\ \text{kN} - 200\ \text{kN} + 400\ \text{kN} - 50\ \text{kN} = 250\ \text{kN}$$

Next, take the moments of all applied loads about A.

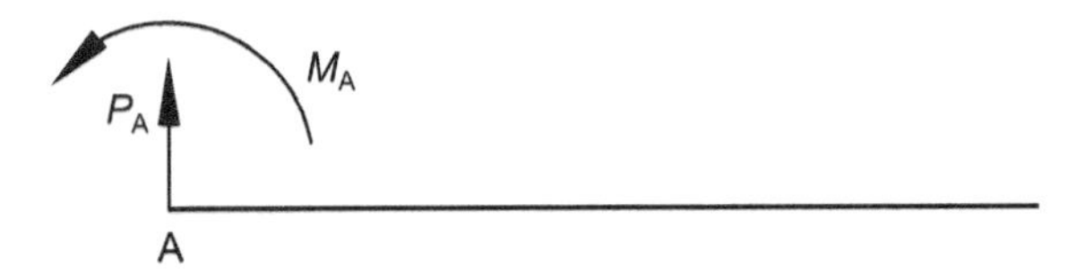

$$\begin{aligned} M_A &= \sum Px \\ &= 100\ \text{kN}(0\ \text{m}) - 200\ \text{kN}(2\ \text{m} + 3\ \text{m}) \\ &\quad +400\ \text{kN}(2\ \text{m} + 3\ \text{m} + 3\ \text{m}) \\ &\quad -50\ \text{kN}(2\ \text{m} + 3\ \text{m} + 3\ \text{m} + 4\ \text{m} + 1\ \text{m}) \\ &= 1550\ \text{kN·m}\ \ (\text{cw}) \end{aligned}$$

The answer is (C).

144. See the *NCEES Handbook*, Statics: Statically Determinate Truss section. The trusses have pinned joints, which can transfer axial forces only. As such, truss members can be subjected to axial loads only.

The answer is (A).

145. The truss tower is a rigid body, and its reactions can be determined using the equilibrium conditions

$$\sum F_x = 0$$
$$\sum F_y = 0$$
$$\sum M = 0$$

Use the sum of moments about B to find V_A.

$$\begin{aligned} \sum M_B &= 0 \\ &= V_A(24\ \text{m}) - (500\ \text{kN})(12\ \text{m}) \\ V_A &= \frac{(500\ \text{kN})(12\ \text{m})}{24\ \text{m}} \\ &= 250\ \text{kN} \end{aligned}$$

Shortcut: The applied vertical load is 500 kN. Therefore, the sum of the vertical reactions at A and B is also 500 kN. Due to symmetry, the vertical reactions at supports A and B are equal. Therefore, both vertical reactions are 250 kN.

The answer is (B).

146. Material properties such as strength and modulus of elasticity are not considered in computing the centroid of a composite section. The centroid can be found by summation.

$$\bar{y} = \frac{\sum y_i A_i}{\sum A_i}$$

The areas of the components are

$$\begin{aligned} A_{\text{beam}} &= 14.1\ \text{in}^2 \\ A_{\text{slab}} &= (6\ \text{in})(60\ \text{in}) = 360\ \text{in}^2 \end{aligned}$$

The centroid of each component is at its middepth.

$$\begin{aligned} y_{\text{beam}} &= \frac{13.8\ \text{in}}{2} = 6.9\ \text{in} \\ y_{\text{slab}} &= 13.8\ \text{in} + \frac{6\ \text{in}}{2} = 16.8\ \text{in} \end{aligned}$$

The distance to the centroid of the composite section is

$$\begin{aligned} \bar{y} &= \frac{\sum y_i A_i}{\sum A_i} \\ &= \frac{(6.9\ \text{in})(14.1\ \text{in}^2) + (16.8\ \text{in})(360\ \text{in}^2)}{14.1\ \text{in}^2 + 360\ \text{in}^2} \\ &= 16.43\ \text{in}\quad (16.4\ \text{in}) \end{aligned}$$

The answer is (D).

147. The MOI of a rectangular section with width b and depth d about its axis is determined using the formula

$$I_1 = \frac{bd^3}{12}$$

For the welded bars, the depth increases to $2d$, as shown.

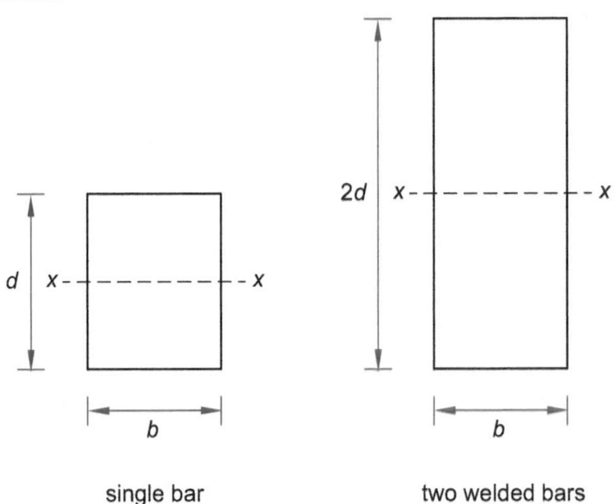

Therefore, the MOI of the welded bars is

$$I_2 = \frac{b(2d)^3}{12} = \frac{8bd^3}{12}$$
$$I_2 = 8I_1$$
$$\frac{I_2}{I_1} = 8$$

The answer is (C).

148. See the *NCEES Handbook*, Statics: Friction section. Limiting friction is defined as the maximum frictional force, which comes into play when a body just begins to slide over the surface of another body. Thus, option A is correct.

The frictional force is a vector and is independent of the area of contact. Thus, statements B and C are incorrect, and statement D is correct.

As a body starts to slide, the force of friction remains constant at moderate speeds, but decreases slightly as the speed increases. Thus, statement E is incorrect.

The answer is (A) and (D).

149. Consider equilibrium along the x-axis and y-axis at impeding motion between the block and the plane, as shown. Determine the force needed to stop the block from sliding down.

$$\begin{aligned}\sum F_y &= 0\\ W &= 500\text{ lbf}\\ N &= W\cos 30° = 500\text{ lbf}(0.866)\\ &= 433\text{ lbf}\\ F_{\text{friction}} &= \mu_s N = 0.5(433\text{ lbf})\\ &= 216.5\text{ lbf}\\ \sum F_x &= 0\\ F_{\text{applied}} + F_{\text{friction}} - W_x &= 0\\ F_{\text{applied}} &= W\sin\theta - F_{\text{friction}}\\ &= (500\text{ lbf})\sin 30° - 216.5\text{ lbf}\\ &= 33.5\text{ lbf}\\ T_1 &\geq 33.5\text{ lbf}\end{aligned}$$

For the pulley, use the belt friction formula.

$$\begin{aligned}\theta &= 90° + 30° = 120°\\ &= \frac{120°\pi}{180} = 2.09\text{ rad}\\ \mu &= 0.2\\ T_2 &= T_1 e^{\mu\theta}\\ &= (33.5\text{ lbf})e^{(0.2)(2.09\text{ rad})}\\ &= 50.88\text{ lbf}\quad(51\text{ lbf})\end{aligned}$$

The answer is (A).

150. Use the truss analysis method given the *NCEES Handbook*. Alternately, use the zero-force member rule, which states that the force in a member is zero if

(a) two noncollinear members are connected to a joint that has no external loads or reactions applied to it; then, the force in both members is zero, or

(b) in a three-member joint, two members are collinear, and the joint has no external loads or reactions applied to it; then, the force in the noncollinear member is zero.

step 1: Since the load is applied to point E, all members (AB, BC, CD, DE, EF) along the bottom chord have nonzero force.

step 2: Since the reactions at supports A and F are nonzero, the forces in the member AJ and FJ are nonzero.

step 3: Apply the zero-force rule to all other members. Thus, the six zero-force members are EK, DJ, CJ, CH, BH, and BG.

The answer is (D).

151. See the *NCEES Handbook*, Dynamics: Work section. Work is defined as the product of force and distance travelled and expressed as

$$U = \int F\cdot dr$$

In defining work, statement A has no displacement, and statement B has no force. Therefore, statements A and B are incorrect. A body can have energy even though it is not moving; it can have energy by virtue of its position. It is called potential energy. Therefore, statement C is correct. In this case, the total energy of a moving body equals its kinetic energy (KE), which is defined as

$$\text{KE} = \frac{m\text{v}^2}{2}$$

In the KE equation, the velocity, v, is constant over time. Therefore, KE also remains constant over time. Statement D is incorrect.

The answer is (C).

152. The question asks for the additional load induced in the front axle. Therefore, the wheel reactions due to gravity load are excluded.

$$a = 0.35\ \text{g}$$
$$m = \frac{W}{g}$$

Consider the energy equation:

$$\begin{aligned} F &= ma \\ &= \frac{10{,}000\ \text{lbf}}{g}(0.35\ \text{g}) = 3500\ \ \text{lbf} \end{aligned}$$

The force F acts horizontally at the CG, which is located 3 ft above ground. Taking the moment about the rear axle B,

$$\begin{aligned} \sum M_{\text{A}} &= 0 \\ V_{\text{A}}(6\ \text{ft} + 3\ \text{ft}) - F(3\ \text{ft}) &= 0 \\ V_{\text{A}} &= \frac{F(3\ \text{ft})}{(6\ \text{ft} + 3\ \text{ft})} \\ &= \frac{(3500\ \text{lbf})(3\ \text{ft})}{9\ \text{ft}} \\ &= 1167\ \text{lbf} \quad (1200\ \text{lbf}) \end{aligned}$$

The answer is (A).

153. See the *NCEES Handbook*, Dynamics: Particle Kinetics section. Use Newton's second law of motion.

$$\sum F_y = ma$$

step 1: Under static condition, for equilibrium, $T = W$, where T is the tension in the cable and W is the total weight of the elevator.

step 2: For dynamic equilibrium, the maximum tension in the cable, T, occurs during the period the elevator is being accelerated upward.

$$\begin{aligned} \sum F_y &= ma \\ T - W &= ma \\ T &= W + ma \\ &= W + \left(\frac{W}{g}\right)a \\ &= 1000\ \text{lbf} + \left(\frac{1000\ \text{lbf}}{32.2\ \frac{\text{ft}}{\text{sec}^2}}\right)\left(10\ \frac{\text{ft}}{\text{sec}^2}\right) \\ &= 1311\ \text{lbf} \quad (1300\ \text{lbf}) \end{aligned}$$

The answer is (D).

154. The applied force, F, is proportional to the spring deformation, x. Therefore, the spring force-deformation relationship is

$$\begin{aligned} F &\propto x \\ F &= kx \end{aligned}$$

k is the spring constant.

$$\begin{aligned} k &= \frac{F}{x} \\ &= \frac{1000\ \ \text{lbf}}{4\ \ \text{in}} = 250\ \frac{\text{lbf}}{\text{in}} \end{aligned}$$

Work done, W, equals force × distance traveled (i.e., from 8 in height to 4 in).

$$\begin{aligned} W &= \int_4^8 F\ dx \\ &= \int_4^8 kx\ dx \\ &= \left(250\ \frac{\text{lbf}}{\text{in}}\right)\left(\frac{x^2}{2}\right)\Bigg|_4^8 \\ &= (6000\ \text{lbf-in})\left(\frac{1\ \text{kip}}{1000\ \text{lbf}}\right) \\ &= 6\ \ \text{in-kip} \end{aligned}$$

The answer is (C).

155. The acceleration, α, of a particle rotating at a distance r from a fixed point has two components: tangential and radial.

$$\alpha_t = r\alpha$$
$$\alpha_r = r\omega^2$$
$$\alpha = \sqrt{\alpha_t^2 + \alpha_r^2}$$

The radial acceleration is

$$\begin{aligned}\alpha_r &= r\omega^2 \\ &= (0.5\ \text{m})\left(20\ \frac{\text{rad}}{\text{s}}\right)^2 \\ &= 200\ \text{m/s}^2\end{aligned}$$

In this case, the shaft is rotating at a constant velocity, so its angular acceleration is zero. Therefore, its tangential acceleration, α_t, is 0.

$$\begin{aligned}\alpha &= \sqrt{\alpha_r^2 + \alpha_t^2} \\ &= \alpha_r \\ &= 200\ \text{m/s}^2\end{aligned}$$

The answer is (B).

156. Kennedy's rule states that when three bodies move relative to one another, they have three instantaneous centers, all of which are on the straight line.

The answer is (A).

157. See the *NCEES Handbook*, Statics: Centroids of Masses, Areas, Lengths, and Volume, Statics: Moment of Inertia, and Dynamics: Mass Moment of Inertia sections.

A body has only one centroid where the entire mass of the body is assumed to be concentrated. Therefore, statement A is incorrect.

The radius of gyration is defined as the distance from a given reference where the whole mass or area of the body is assumed to be concentrated to give the same value of the moment of inertia. Statement B is incorrect.

The least moment of inertia is at the centroidal axis of the section. Therefore, statement C is incorrect, and statement D is correct.

The answer is (D).

158. According to the *NCEES Handbook*, Materials Science/Structure of Matter section, the Charpy test is used to determine the amount of energy required to cause failure in standardized specimen. The principle of a Charpy machine is that the pendulum's potential energy is lost in causing failure of the test specimen. Use the *NCEES Handbook*, Dynamics: Potential Energy section to determine the energy. The energy absorbed by the specimen in breaking, ΔU, is equal to the potential energy (PE) at the angle at which the pendulum was released (point A), minus the PE at the angle the pendulum reached after causing failure of the specimen (point B).

step 1: Take the point B as the reference plane. The general equation for the height of the pendulum's center of mass above the reference plane for any angle θ is

$$h = R(1 - \cos\theta)$$

step 2: Determine the initial energy of the pendulum.

$$\begin{aligned}U_\text{A} &= mg(h_\text{A}) \\ &= WR(1 - \cos\alpha)\end{aligned}$$

step 3: Determine the pendulum energy after breaking.

$$\begin{aligned}U_\text{B} &= mg(h_\text{B}) \\ &= WR(1 - \cos\beta)\end{aligned}$$

step 4: Determine the loss of pendulum energy in causing failure of the specimen.

$$\begin{aligned}\Delta U &= U_\text{A} - U_\text{B} \\ &= WR(1 - \cos\alpha) - WR(1 - \cos\beta) \\ &= WR(\cos\beta - \cos\alpha) \\ &= (10\ \text{lbf})(3\ \text{ft})(\cos 15° - \cos 90°) \\ &= 28.98\ \text{ft-lbf} \quad (29\ \text{ft-lbf})\end{aligned}$$

The answer is (C).

159. In this case, the car's engine provides the kinetic energy, which is dissipated by the car frontal end getting crushed, and the car length shortens. This is the distance needed to bring the car speed to zero. See the *NCEES Handbook*, Dynamics: Kinetic Energy and Work sections. Use conservation of energy and the definition of work done.

$$\begin{aligned}\text{Kinetic energy of car} &= \text{Work done} \\ &= \text{Impact Force} \\ &\quad \times \text{distance travelled by the force}\end{aligned}$$

$$\frac{mv^2}{2} = F \times \delta$$

$$F = \frac{mv^2}{2\delta} = \left(\frac{W}{g}\right)\left(\frac{v^2}{2\delta}\right) = \frac{Wv^2}{2g\delta}$$

step 1: Convert the car's speed from mph to ft/sec, using the *NCEES Handbook*, Units and Conversion Factors section.

$$v = \left(30\ \frac{\text{mi}}{\text{hr}}\right)\left(88.0\ \frac{\frac{\text{ft}}{\text{min}}}{\frac{\text{mi}}{\text{hr}}}\right)\left(\frac{1\ \text{min}}{60\ \text{sec}}\right) = 44.0\ \text{ft/sec}$$

step 2: Determine the distance, δ.

$$\delta = \Delta L = L_1 - L_2 = 2\ \text{ft}$$

step 3: Determine the impact force on the car.

$$F = \frac{Wv^2}{2g\delta} = \frac{\left((6000\ \text{lbf})\left(\frac{1\ \text{kip}}{1000\ \text{lbf}}\right)\right)(44.0\ \text{ft/sec})^2}{(2)(32.174\ \text{ft/sec}^2)(2\ \text{ft})} = 90.26\ \text{kips}\quad (91\ \text{kips})$$

The answer is (C).

160. See *NCEES Reference Handbook*, Dynamics: Free and Forced Vibrations. The equation of motion for a single degree-of-freedom vibration system, containing a mass m, a spring k, and a viscous damper c, in terms of y, is

$$m\ddot{y} + c\dot{y} + ky = 0$$

step 1: For an undamped system, damping, $c = 0$. Therefore, the equation reduces to

$$m\ddot{y} + ky = 0$$

$$k = 8\ \frac{\text{lbf}}{\text{ft}}$$

$$m = \frac{W}{g} = \frac{4\ \text{lbf}}{32\ \text{ft}^2/\text{sec}} = \frac{1}{8}\ \text{lbf-sec/ft}^2$$

Substitute the values of k and m into the differential equation.

$$\frac{1}{8}\ddot{y} + 8y = 0$$

$$\ddot{y} + 64y = 0$$

The answer is (A).

161. The maximum bending moment occurs at the section where shear force is zero. In this case, the shear force is zero at point C. Therefore, the maximum bending moment is at point C. The change in the bending moment between two points in a beam equals the area of the S.F. diagram between the two points.

$$M_C - M_A = \text{Area of S.F. diagram between points A and C}$$

Since the beam is simply supported,

$$M_A = 0$$

$$M_C = \text{S.F. diag. area between pts. A and C} = (2400\ \text{kN})(6\ \text{m}) + (1400\ \text{kN})(4\ \text{m}) = 20{,}000\ \text{kN·m}$$

The answer is (C).

162. Use the Mohr Circle method as given in the *NCEES Handbook*. Draw the circle as shown.

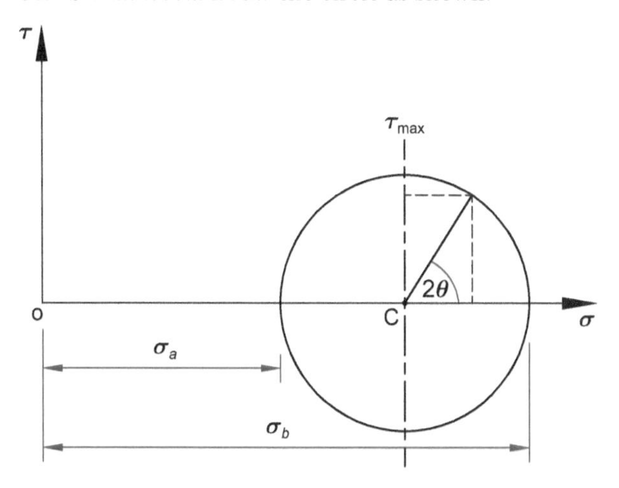

The circle is drawn with the center on the normal stress. Its center C is located at

$$C = \frac{\sigma_x + \sigma_y}{2} = \frac{100 \text{ ksi} + 50 \text{ ksi}}{2} = 75 \text{ ksi}$$

The radius of the Mohr's circle is

$$R = \sqrt{\left(\frac{\sigma_x - \sigma_y}{2}\right)^2 + \tau_{xy}^{\;2}} = \sqrt{\left(\frac{100 \text{ ksi} - 50 \text{ ksi}}{2}\right)^2 + 0^2} = 25 \text{ ksi}$$

The answer is (A).

163. This problem describes a uniaxial case of Hooke's law. The bar under tension is shown.

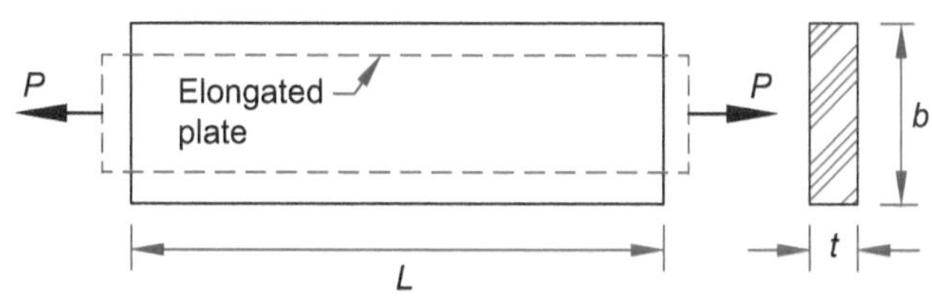

As the bar elongates, it decreases in width. The decrease in width is a function of lateral strain and the plate width.

$$\text{Change in width} = \epsilon_y b$$

The material properties of aluminum are given in the Typical Material Properties table in the Mechanics of Materials section of the *NCEES Handbook*. From the conversion factors table in the Units and Conversion Factors section of the *NCEES Handbook*

$$1 \text{ Pa} = 1 \ \frac{\text{N}}{\text{m}^2}$$
$$1 \text{ kPa} = 1 \ \frac{\text{kN}}{\text{m}^2}$$

step 1: Determine the Young's modulus of aluminum, as given in the *NCEES Handbook.*

$$E = 69 \text{ GPa} = 69 \times 10^9 \text{ Pa} = 69 \times 10^9 \ \frac{\text{N}}{\text{m}^2}$$

step 2: Determine cross-sectional area of the bar.

$$A = bt = (20 \text{ cm})(20 \text{ mm})\left(\frac{1 \text{ cm}}{10 \text{ mm}}\right) = 40 \text{ cm}^2 = 4 \times 10^{-3} \ \text{m}^2$$

step 3: Determine axial stress in the bar.

$$\sigma = \frac{P}{A} = \frac{300 \text{ kN}}{4 \times 10^{-3} \ \text{m}^2} = 75{,}000 \ \frac{\text{kN}}{\text{m}^2} = 75 \times 10^6 \ \frac{\text{N}}{\text{m}^2}$$

step 4: Determine the longitudinal strain.

$$\epsilon_x = \frac{\sigma}{E} = \frac{75 \times 10^6 \text{ N/m}^2}{69 \times 10^9 \text{ N/m}^2} = 1.087 \times 10^{-3}$$

step 5: Determine the lateral strain. The *NCEES Handbook*, Mechanics of Materials: Material Properties section, gives Poisson's ratio, ν, for aluminum as 0.33.

The lateral strain is

$$\epsilon_y = \nu \ \epsilon_x = 0.33 \ (1.087 \times 10^{-3}) = 0.3587 \times 10^{-3}$$

step 6: Determine decrease in width.

$$\epsilon_y b = (0.3587 \times 10^{-3})0.2 \text{ m} = 71.74\ \mu\text{m} \quad (72\ \mu\text{m})$$

The answer is (C).

164. The bars are solid with a rectangular shape. Their stiffness is proportional to the moment of inertia (MOI) of the section. For a rectangular section, the MOI is given by

$$I = \frac{bd^3}{12}$$

The two bars placed one on the top of the other act independently and noncompositely. Both carry equal loads, and both deflect equally, as shown.

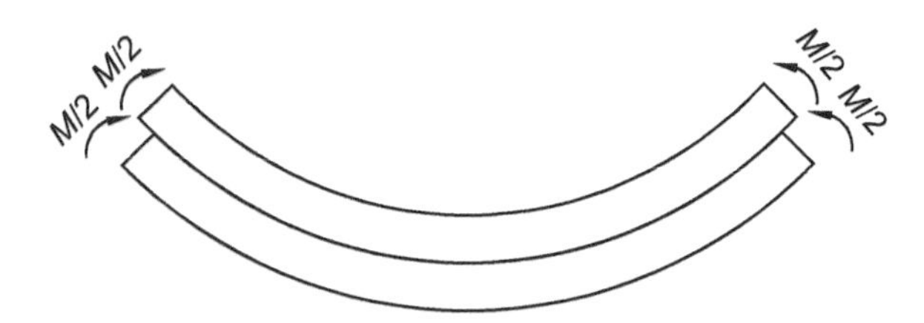

$$I_{\text{noncomp}} = \sum I = \frac{bd^3}{12} + \frac{bd^3}{12} = (2)\left(\frac{bd^3}{12}\right)$$

After welding, the two sections act as a composite section that has a total depth of $2d$. Its MOI is given by

$$I_{\text{comp}} = \frac{b(2d)^3}{12} = \frac{8bd^3}{12} = (4)\left(\frac{2bd^3}{12}\right) = 4\ (I_{\text{noncomp}})$$

The ratio of composite MOI to the noncomposite MOI is 8/2, or 4:1.

The answer is (B).

165. The angle of rotation of the tube, in radians, is given by

$$\phi = \frac{TL}{GJ}$$

The shear modulus, G, for steel is 11 Mpsi.

step 1: Determine the polar moment of inertia, J, for a hollow tube.

$$J = \frac{\pi({d_2}^4 - {d_1}^4)}{32}$$

The outer diameter is

$$d_2 = 12 \text{ in}$$

The inner diameter is

$$d_1 = 12 \text{ in} - (2)(1 \text{ in}) = 10 \text{ in}$$

The polar moment of inertia is

$$J = \frac{\pi({d_2}^4 - {d_1}^4)}{32} = \frac{\pi\left((12 \text{ in})^4 - (10 \text{ in})^4\right)}{32} = 1053 \text{ in}^4$$

Use this value for the polar moment of inertia to calculate the angle of rotation of the tube.

$$\phi = \frac{TL}{GJ} = \frac{\left((100 \text{ ft-kips})\left(12\ \frac{\text{in}}{\text{ft}}\right)\right)\left((10 \text{ ft})\left(12\ \frac{\text{in}}{\text{ft}}\right)\right)}{\left(11 \times 10^3\ \frac{\text{kips}}{\text{in}^2}\right)(1053 \text{ in}^4)} = \frac{0.01243 \text{ rad } (180^\circ)}{\pi} = 0.7122^\circ \quad (0.71^\circ)$$

The answer is (B).

166. See the *NCEES Handbook*, Mechanics of Materials section.

The longitudinal strain is defined as the elongation of a bar per unit length in the direction of the force. Therefore, statement A is incorrect.

Poisson's ratio is defined as the ratio of lateral strain to longitudinal strain. Therefore, statement B is incorrect.

When a body is subjected to a triaxial loading of equal intensity, the ratio of direct stress to the corresponding strain is called bulk modulus. Therefore, statement C is incorrect.

The volumetric strain is defined as the ratio of change in volume to original volume. Therefore, statement D is correct.

The answer is (D).

167. As the temperature rises, bar A tends to expand more than bar B, as shown.

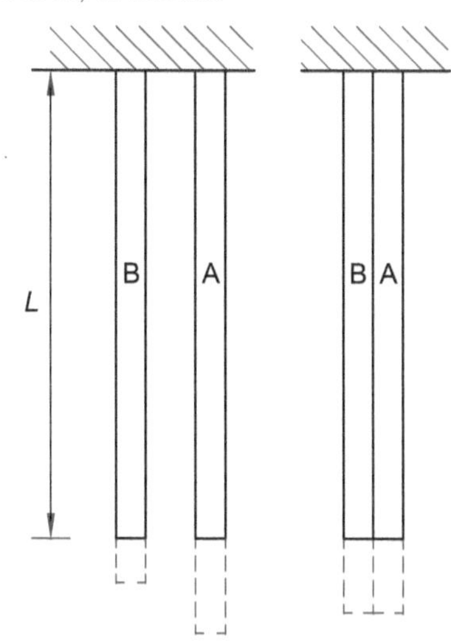

In an unrestrained situation, neither bar would have any induced stress. If they are restrained to expand or contract equally, stresses would be induced in both bars. Bar A is not able to fully expand due to restraint caused by bar B. As such, bar A is in compression, bar B is in tension, and both bars expand equally.

The answer is (A).

168. See the *NCEES Handbook*, Mechanical Engineering section. The combined shear and moment load is resisted by the four-bolt assembly. The applied system of forces on bolt 1 is shown.

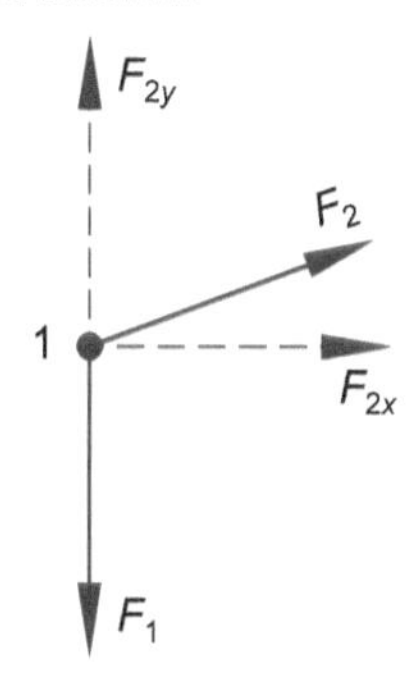

Determine the resultant of applied forces F_1 and F_2. The forces F_1 and F_2 are shown in the *NCEES Handbook* in its fastener groups force diagram. Consider downward force as negative.

step 1: The magnitude of direct shear force on each bolt due to P is

$$F_1 = \frac{P}{n} = \frac{-24 \text{ kips}}{4} = -6 \text{ kips}$$

The force acts in the same direction to the applied load, P.

step 2: Resolve the applied bending moment to the system of shear forces on the bolts. The magnitude of shear force due to moment M is

$$F_2 = \frac{M \times r_1}{\sum_{i=1}^{n} r_i^2}$$

For convenience, resolve the force F_2 in the x- and y-directions.

$$F_{2x} = \frac{M \times y_1}{\sum_{i=1}^{4} r_i^2}$$

$$F_{2y} = \frac{M \times x_1}{\sum_{i=1}^{4} r_i^2}$$

Determine distance r between the bolts and the centroid using the table given in the problem statement.

$$r^2 = x^2 + y^2$$

$$\begin{aligned}\sum r^2 &= \sum x^2 + \sum y^2 \\ &= 36 \text{ in}^2 + 16 \text{ in}^2 \\ &= 52 \text{ in}^2\end{aligned}$$

The directions of the forces are shown.

$$\begin{aligned}F_{2x} &= \frac{M \times y}{\sum_{i=1}^{4} r_i^2} \\ &= \frac{(312 \text{ in-kips})(2 \text{ in})}{52 \text{ in}^2} = 12 \text{ kips}\end{aligned}$$

$$\begin{aligned}F_{2y} &= \frac{M \times x}{\sum_{i=1}^{4} r_i^2} \\ &= \frac{(312 \text{ in-kips})(3 \text{ in})}{52 \text{ in}^2} = 18 \text{ kips}\end{aligned}$$

step 3: Determine the resultant force using the vector method given in the *NCEES Handbook*, Statics section. The total shear force on a bolt is the vector sum of the two forces.

$$\begin{aligned}\sum F_x &= F_{1x} + F_{2x}\\ &= 0 \text{ kips} + 12 \text{ kips}\\ &= 12 \text{ kips}\end{aligned}$$

$$\begin{aligned}\sum F_y &= F_{1y} + F_{2y}\\ &= -6 \text{ kips} + 18 \text{ kips}\\ &= 12 \text{ kips}\end{aligned}$$

The resultant shear on bolt 1 is

$$\begin{aligned}F_{\text{total}} &= \sqrt{(12 \text{ kips})^2 + (12 \text{ kips})^2}\\ &= 12 \text{ kips}\end{aligned}$$

The answer is (B).

169. The shaft is carrying a UDL of 2.5 kN/m (2500 N/m). For deflection, see the *NCEES Handbook*, Mechanics of Materials: Simply Supported Beam Slopes and Deflections table.

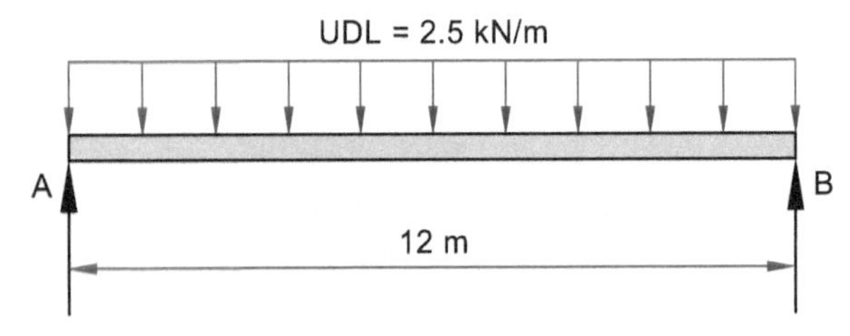

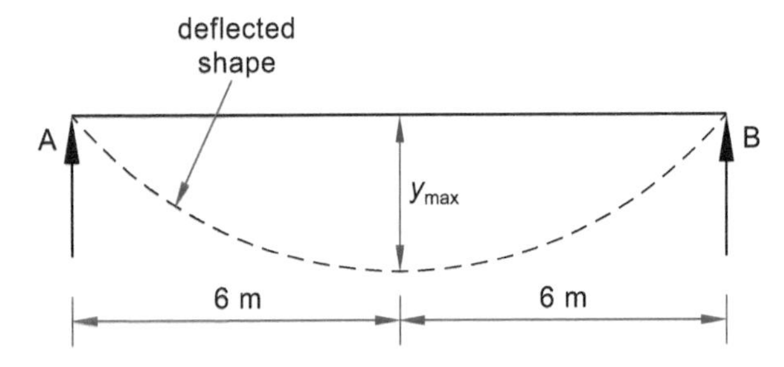

The maximum deflection occurs at its midspan.

$$y_{\text{max}} = \frac{5wL^4}{384EI}$$

step 1: Determine the moment of inertia of the hollow shaft.

$$I = \frac{\pi(a^4 - b^4)}{4}$$

The outer radius is

$$\begin{aligned}a &= \frac{d}{2}\\ &= \frac{400 \text{ mm}}{2}\\ &= 200 \text{ mm}\\ &= 0.2 \text{ m}\end{aligned}$$

The inner radius is

$$\begin{aligned}b &= a - t_{\text{pipe}}\\ &= 200 \text{ mm} - 20 \text{ mm}\\ &= 180 \text{ mm}\\ &= 0.18 \text{ m}\end{aligned}$$

The modulus of elasticity is

$$\begin{aligned}I &= \frac{\pi\left((0.2 \text{ m})^4 - (0.18 \text{ m})^4\right)}{4}\\ &= 0.0004322 \text{ m}^4\end{aligned}$$

step 2: Determine steel's modulus of elasticity.

$$\begin{aligned}E &= 200 \text{ GPa}\\ &= 200 \times 10^9 \text{ Pa}\\ &= 200 \times 10^9 \ \frac{\text{N}}{\text{m}^2}\end{aligned}$$

The maximum deflection is

$$\begin{aligned}y_{\text{max}} &= \frac{5\left(2500 \ \frac{\text{N}}{\text{m}}\right)(12 \text{ m})^4}{384\left(200 \times 10^9 \ \frac{\text{N}}{\text{m}^2}\right)(0.0004322 \text{ m}^4)}\\ &= 0.007809 \text{ m}\\ &= 7.809 \text{ mm} \quad (7.8 \text{ mm})\end{aligned}$$

The answer is (B).

170. The W-shape properties are given the *NCEES Handbook*, Design of Steel Components section, W Shapes Dimensions and Properties table. The column buckling capacity, P_{cr}, can be determined using the Euler equation.

$$P_{\text{cr}} = \frac{\pi^2 EI}{(KL)^2}$$

For a W14×74, $I_x = 795 \text{ in}^4$ and $I_y = 134 \text{ in}^4$.

The column has pinned conditions at both ends. Since the column would buckle about the weaker axis, I_y governs.

$$E = 29{,}000 \text{ ksi}$$
$$L = (20 \text{ ft})\left(12 \ \frac{\text{in}}{\text{ft}}\right) = 240 \text{ in}$$

For a pinned-pinned column, $K = 1$.

$$P_{cr} = \frac{\pi^2\left(29{,}000 \ \frac{\text{kips}}{\text{in}^2}\right)(134 \text{ in}^4)}{\left((1)(240 \text{ in})\right)^2} = 665.9 \text{ kips} \quad (670 \text{ kips})$$

The answer is (B).

171. See the *NCEES Handbook*, Civil Engineering: Structural Analysis section. The beam with the applied moment, M_A, is shown.

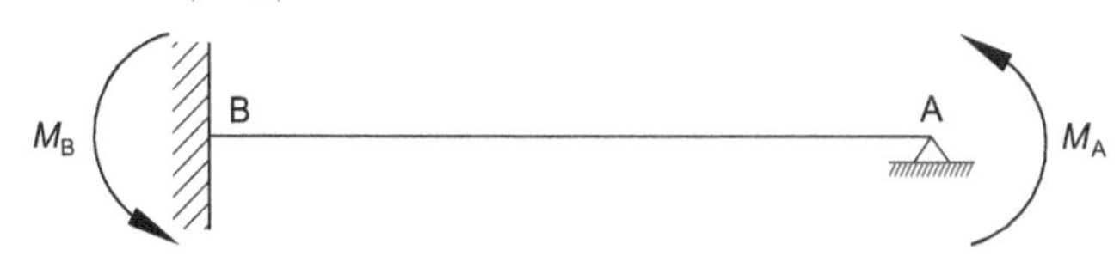

Support B is the fixed end of the propped cantilever. The induced moment at support B, M_B, is called the carryover moment. The carryover factor is expressed as the ratio

$$\text{Carryover factor} = \frac{M_B}{M_A}$$

The associated carryover factor is one-half. Option B is correct.

The answer is (B).

172. See the *NCEES Handbook*, Materials Science section. The tensile strength of a material is related to its hardness.

$$\begin{aligned}\text{Tensile strength} &= 500 \text{ BHN psi}\\ &= 500(98) \text{ psi}\\ &= 49{,}000 \text{ psi}\\ &= 49 \text{ ksi}\end{aligned}$$

The answer is (B).

173. The stress is defined as force per unit area. It is also defined as

$$\text{stress, } \sigma = \text{strain} \times \text{Modulus of elasticity}$$

The stress at which a material will experience permanent deformation is called the yield stress. The yield stress is the onset when the material is no longer elastic and undergoes a permanent set. It is shown as the yield point.

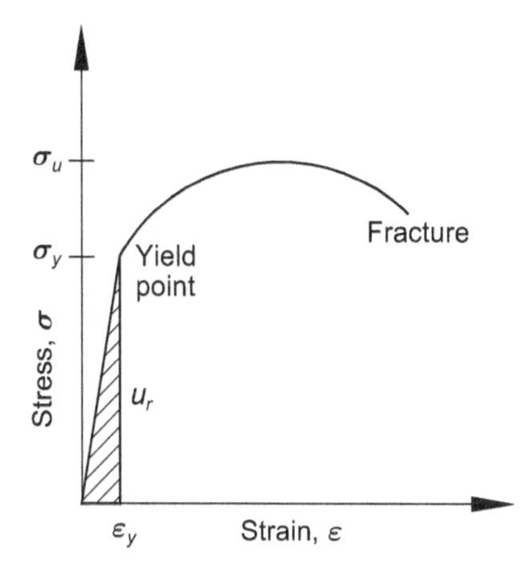

The answer is (A).

174. See the *NCEES Handbook*, Materials Science/Structure of Matter section. Fracture toughness depends on the applied stress and the crack length. The stress intensity at which the material will fail is

$$K_{IC} = Y\sigma\sqrt{\pi a}$$

It is given that $Y = 1$. Therefore, maximum applied stress is

$$\sigma = \frac{K_{IC}}{\sqrt{\pi a}}$$

step 1: For central crack, crack length $= 2a$. Therefore,

$$a = \frac{0.58 \text{ in}}{2} = 0.29 \text{ in}$$

step 2: For the plate material,

$$K_{IC} = 22 \text{ (ksi-in)}^{1/2}$$

step 3: Apply the formula and determine critical stress. Therefore,

$$\sigma = \frac{22 \text{ (ksi-in)}^{1/2}}{\sqrt{\pi(0.29 \text{ in})}} = 23 \text{ ksi}$$

The answer is (A).

175. See the *NCEES Handbook*, Materials Science/Structure of Matter section. Hardness is resistance to penetration; hardenability is the "ease" with which hardness can be obtained.

Hardness, and not hardenability, is a measure of resistance of plastic deformation as measured by indentation. Hardenability is the ability or potential of a steel to achieve a certain hardness at a given depth, upon suitable heat treatment and quench. The two terms are not synonymous. Therefore, the statements A and B are incorrect.

Hardenability is not for all metals or their alloys. Hardness can be measured in metals in any condition. Hardenability presumes that the steels will be heat-treated to achieve a targeted hardness at a given depth. Therefore, statement C is incorrect.

It is true that hardenability can be gauged by using the Jominy (H-band) or curves. The hardenability varies with the steel chemistry and the quenching rates, as shown in the *NCEES Handbook* graph.

The answer is (D).

176. The cold working process reduces toughness, impact strength, and ductility. Therefore, statement A is incorrect. The cold working improves hardness. Therefore, statement B is incorrect.

The cold work (CW) is a measure of the degree of plastic deformation. For example, cold work done to reduce the crosssection of a rod can be calculated as follows:

$$\text{CW done} = \frac{\begin{array}{c}\text{Area of cross section before CW}\\ -\text{Area of cross section after CW}\end{array}}{\text{Area of cross section before CW}}$$

CW can also be expressed in percentage terms. Therefore, statement C is incorrect.

The cold working process produces good surface finish on the metal. Therefore, statement D is correct.

The answer is (D).

177. See the *NCEES Handbook*, Chemistry and Biology Oxidation Potentials for Corrosion Reactions table.

It is true that corrosion is a chemical reaction. In the corrosion process, there is a transfer of electrons from one chemical species to another. Therefore, option A is a correct statement.

It is true that there is no method of measuring the absolute value of an electrode potential. All electrode potentials are determined under a standard condition with reference to a standard hydrogen electrode whose value is considered as zero, as shown in the table. Therefore, option B is a correct statement.

The metals having electrode potential lower than that of hydrogen are known as anodic metals, and not as cathodic metals. Option C is a correct statement.

The surface of the object changes from an element to a compound and not to another element. For example, aluminum develops a thin oxidation layer immediately upon exposure to the atmosphere. The oxide film protects the surface from further oxidation. This is a change from an element to a compound. Therefore, option D is an incorrect statement.

The answer is (D).

178. A relatively large range of rotary motion is required at the hip by a ball-and-socket type of joint. The joint is susceptible to fatigue and fracture. Diseased and fractured joint have been successfully replaced by a metallic joint. The problem statement isolates the fatigue and fracture as the sole criterion for selection.

The term "fatigue fracture" is defined as the fracture that is caused by repeatedly applied fatigue stresses. The fatigue strength is directly related to the ultimate strength of the metal. The fatigue fracture stresses are well below the tensile strength of the metal or alloy used. The higher the tensile strength, the higher the fatigue fracture stress, given the fatigue stress level.

step 1: Determine the ultimate strength of the typical engineering materials: See the *NCEES Handbook*, Mechanics of Materials: Average Mechanical Properties of Typical Engineering Materials table. Select materials that have the highest yield stress and ultimate strength, as shown:

Material	Yield stress, σ_y (ksi)	Ultimate strength, σ_u (ksi)
A. Stainless steel 304	30	75
B. Aluminum 2014-T6	60	68
C. Titanium alloy (Ti-6Al-4V)	134	145
D. Bronze C86100	50	95

step 2: Select the material with the highest ultimate strength. It is titanium alloy (Ti-6Al-4V).

The answer is titanium alloy.

179. See the *NCEES Handbook*, Units and Conversion Factors section.

$$1 \text{ psi} = 0.068 \text{ atm}$$

$$\begin{aligned}\text{Tire pressure} &= 29.5 \text{ psi}\left(\frac{0.068 \text{ atm}}{1 \text{ psi}}\right)\\ &= 2.01 \text{ atm} \quad (2.0 \text{ atm})\end{aligned}$$

The answer is (C).

180. The dam acts as a retaining wall to resist the water lateral pressure. See the *NCEES Handbook*, Fluid Mechanics section, for horizontal stress profiles and force for active forces on retaining wall per unit wall length. The total lateral force caused by the hydrostatic pressure at base of the dam is also called a sliding force.

The maximum hydrostatic pressure at depth, h, from the top is

$$p_h = \gamma_{\text{water}}(h)$$

In this equation, γ_{water} is the specific weight of the water, or 62.4 lbf/ft^3, and h is the height of the water column. The figure has the corresponding lateral force per unit length.

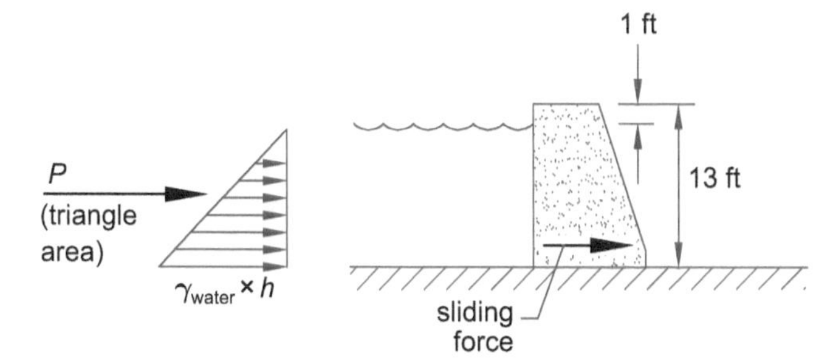

The corresponding hydrostatic force per unit width of the wall is

$$P_h = \frac{1}{2}(\gamma_{\text{water}})(h^2)$$

The force equals the area under the hydrostatic pressure diagram. The maximum lateral force is at the base, is shown in the figure. The water height above the base is

$$h = 13 \text{ ft} - 1 \text{ ft} = 12 \text{ ft}$$

The pressure at the base is

$$\begin{aligned} P_{\text{base}} &= \left(\frac{1}{2}\right)\left(62.4 \ \frac{\text{lbf}}{\text{ft}^3}\right)(12 \text{ ft})^2 \\ &= 4493 \text{ lbf/ft} \end{aligned}$$

For the 100 ft length of the dam, the total lateral pressure at the base is

$$\begin{aligned} P_{\text{total}} &= (100 \text{ ft})\left(4493 \ \frac{\text{lbf}}{\text{ft}}\right) \\ &= 449{,}300 \text{ lbf} \\ &= 449.3 \text{ kips} \quad (449 \text{ kips}) \end{aligned}$$

The answer is 449 kips.

181. See the *NCEES Handbook*, (1) Fluid Mechanics section for formulas and (2) Units and Conversion Factors section. Two equations are given in the *NCEES Handbook*, one equation for the force exerted on the plate and the other for work done. For a moving plate, work done is

$$\begin{aligned} \dot{W} &= Q\rho(\text{v}_1 - \text{v})(1 - \cos\alpha)\text{v} \\ \text{v} &= 50 \text{ ft/sec} \\ \text{v}_1 &= 300 \text{ ft/sec} \end{aligned}$$

It is given that the plate moves in the direction of the jet. Therefore, the angle α is 180°. Since cos 180° is zero, the work equation becomes

$$\begin{aligned} \dot{W} &= Q\rho(\text{v}_1 - \text{v})(1 - 0)\text{v} \\ &= Q\rho(\text{v}_1 - \text{v})\text{v} \\ &= \dot{m}(\text{v}_1 - \text{v})\text{v} \end{aligned}$$

step 1: Determine the jet area.

$$\begin{aligned} A &= \frac{\pi}{4}d^2 \\ &= \frac{\pi}{4}\left((1 \text{ in})\left(\frac{1 \text{ ft}}{12 \text{ in}}\right)\right)^2 \\ &= 0.005454 \text{ ft}^2 \end{aligned}$$

step 2: Determine jet mass flow rate hitting the moving plate.

$$\begin{aligned} \dot{m} &= \rho A \text{v} \\ &= \left(1.940 \ \frac{\text{lbf-sec}^2}{\text{ft}^4}\right)(0.005454 \text{ ft}^2) \\ &\quad \times (300 \text{ ft/sec} - 50 \text{ ft/sec}) \\ &= 2.642 \text{ lbf-sec/ft} \end{aligned}$$

step 3: Apply the work-done formula.

$$\begin{aligned} \dot{W} &= \dot{m}(\text{v}_1 - \text{v})\text{v} \\ &= \left(2.642 \ \frac{\text{lbf-sec}}{\text{ft}}\right)\left(300 \ \frac{\text{ft}}{\text{sec}} - 50 \ \frac{\text{ft}}{\text{sec}}\right) \\ &\quad \times \left(50 \ \frac{\text{ft}}{\text{sec}}\right)\left(\frac{1 \text{ hp}}{550 \ \frac{\text{ft-lbf}}{\text{sec}}}\right) \\ &= 60.05 \text{ hp} \quad (60 \text{ hp}) \end{aligned}$$

The answer is (B).

182. See the *NCEES Handbook* for: (1) Fluid Mechanics section for formulas, and (2) the Units and Conversion Factors section. The head loss in the pipe depends on the flow's Reynolds number.

$$\text{Re} = \frac{\text{v}d}{\nu}$$

step 1: Determine oil flow velocity. The pipe area is

$$\begin{aligned} A &= \frac{\pi}{4}d^2 \\ &= \frac{\pi}{4}\left((6\ \text{in})\left(\frac{1\ \text{ft}}{12\ \text{in}}\right)\right)^2 \\ &= 0.1963\ \text{ft}^2 \end{aligned}$$

Determine the discharge, Q, from the known oil weight.

$$\begin{aligned} \dot{m} &= \rho Q \\ Q &= \frac{\dot{m}}{\rho} \\ &= \frac{\left(22\ \frac{\text{tons}}{\text{hr}}\right)\left(2240\ \frac{\text{lbf}}{\text{ton}}\right)\left(\frac{1\ \text{hr}}{3600\ \text{sec}}\right)}{57\ \frac{\text{lbf}}{\text{ft}^3}} \\ &= 0.2402\ \text{ft}^3/\text{sec} \end{aligned}$$

Determine the oil flow velocity.

$$\begin{aligned} \text{v} &= \frac{Q}{A} \\ &= \frac{0.2402\ \frac{\text{ft}^3}{\text{sec}}}{0.1963\ \text{ft}^2} \\ &= 1.224\ \text{ft/sec} \end{aligned}$$

step 2: Determine the Reynolds number.

$$\begin{aligned} \text{Re} &= \frac{\text{v}d}{\nu} \\ &= \frac{\left(1.224\ \frac{\text{ft}}{\text{sec}}\right)\left((6\ \text{in})\left(\frac{1\ \text{ft}}{12\ \text{in}}\right)\right)}{0.02\ \frac{\text{ft}^2}{\text{sec}}} \\ &= 30.6 \quad (31) \end{aligned}$$

Since the Reynolds number is less than 2000, the oil flow is laminar.

The answer is (A).

183. Calculate the Reynolds number for the pole. The wind velocity is

$$\begin{aligned} \text{v} &= \left(65\ \frac{\text{km}}{\text{h}}\right)\left(1000\ \frac{\text{m}}{\text{km}}\right)\left(\frac{1\ \text{h}}{3600\ \text{s}}\right) \\ &= 18.06\ \text{m/s} \end{aligned}$$

The Reynolds number is

$$\begin{aligned} \text{Re} &= \frac{\text{v}d}{\nu} = \frac{\left(18.06\ \frac{\text{m}}{\text{s}}\right)(0.2\ \text{m})}{1.47\times 10^{-5}\ \frac{\text{m}^2}{\text{s}}} \\ &= 2.457\times 10^5 \quad (250{,}000) \end{aligned}$$

Use the Drag Coefficient for Spheres, Disks, and Cylinders figure given in the Fluid Mechanics section of the *NCEES Handbook* to find the drag coefficient for a cylinder with a Reynolds number of 250,000. The corresponding factor is 1.0.

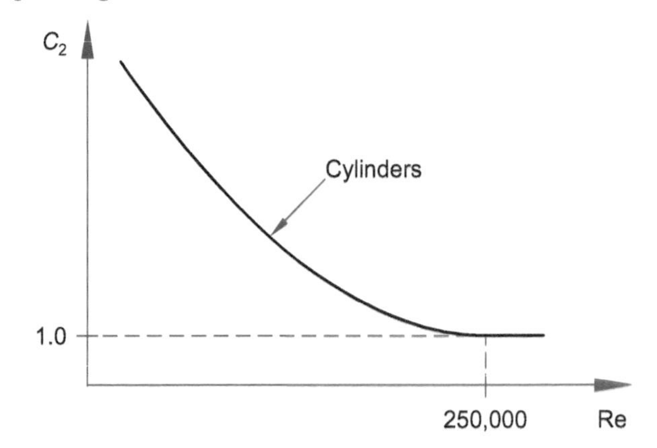

The answer is (B).

184. In a convergent nozzle, the cross-sectional area of the nozzle decreases continuously from entrance to exit as shown.

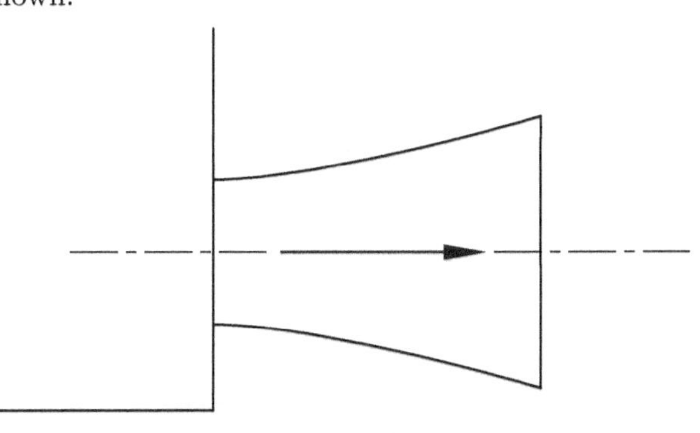

The mass of the fluid passing through the nozzle remains constant, and the fluid's heat energy is converted to kinetic energy with some loss. In an isentropic process, the loss is neglected. See the *NCEES Handbook*: (1) Fluid Mechanics section for energy formula, (2) Thermodynamics section for steam tables, and (3) the

Units and Conversion Factors section. In an ideal gas for an isentropic process, the energy relationship (per unit mass) between the entry and exit points is

$$h_i + \frac{v_i^2}{2} = h_e + \frac{v_e^2}{2}$$

step 1: It is known that the steam velocity at the entrance is negligible. Therefore,

$$\begin{aligned} v_i &= 0 \\ h_i - h_e &= \frac{v_e^2}{2} - \frac{v_i^2}{2} \\ \Delta h &= \frac{v_e^2}{2} \\ v_e &= \sqrt{2\Delta h} \end{aligned}$$

step 2: The nozzle pressure at the inlet is known.

$$p_1 = 10 \text{ bar} = 1 \text{ MPa}$$

From steam tables, the corresponding pressure, enthalpy is

$$h_i = 2778.1 \text{ kJ/kg}$$

At the nozzle exit, the pressure is

$$p_e = 1 \text{ bar} = 0.1 \text{ MPa}$$

From steam tables, the corresponding pressure, enthalpy is

$$h_e = 2675.5 \text{ kJ/kg}$$

step 3: Determine the enthalpy drop in Joule-units.

$$\begin{aligned} \Delta h &= h_i - h_e \\ &= 2778.1 \text{ kJ/kg} - 2675.5 \text{ kJ/kg} \\ &= 102.6 \text{ kJ/kg} \\ &= 102.6 \times 10^3 \text{ J/kg} \end{aligned}$$

step 4: Determine the velocity of steam leaving the nozzle.

$$\begin{aligned} v_e &= \sqrt{(2)(102{,}600 \text{ J/kg})} \\ &= 453.0 \text{ m/s} \quad (453 \text{ m/s}) \end{aligned}$$

The answer is (D).

185. For the frictional resistance or head loss for fluid flowing a segment of pipeline, use the Darcy-Weisbach equation, as given in the given in the *NCEES Handbook*, Fluid Mechanics section.

$$h_f = f\frac{L}{D}\frac{v^2}{2g}$$

step 1: Express the equation in terms of fluid discharge, Q.

$$\begin{aligned} Q &= Av \\ v &= \frac{Q}{A} = \frac{4Q}{\pi D^2} \end{aligned}$$

step 2: Use the above velocity expression in the Darcy-Weisbach equation for a single pump operating.

$$\begin{aligned} h_{f1} &= f\frac{L}{D}\frac{v^2}{2g} = f\frac{L}{D}\frac{\left(\frac{4Q_1}{\pi D^2}\right)^2}{2g} \\ &= f\frac{L}{D^5}\frac{Q_1^2}{2g}\left(\frac{4}{\pi}\right)^2 \\ &= \left(\frac{1}{2g}\left(\frac{4}{\pi}\right)^2\right)f\frac{LQ_1^2}{D^5} \\ &= kf\frac{LQ_1^2}{D^5} \end{aligned}$$

step 3: With two pumps in parallel, the discharge is doubled while the pipe size remains the same. The new discharge is $Q_2 = 2Q_1$.

The head loss for two pumps is

$$\begin{aligned} h_{f2} &= kf\frac{L(2Q_1)^2}{D^5} \\ &= 4\left(kf\frac{LQ_1^2}{D^5}\right) \\ &= 4h_{f1} \end{aligned}$$

Therefore,

$$\frac{h_{f2}}{h_{f1}} = 4$$

The answer is (D).

186. The curves for pumps with 1750 rpm are given in the *NCEES Handbook*, Fluid Mechanics section.

step 1: Locate the curve (C) for the 6 3/4 in diameter pumps.

step 2: The x-axis shows discharge rates. Locate $Q =$ 80 gpm and read vertically to the pump curve.

step 3: The remaining parameters are read at this point. At a capacity of 80 gpm, the curve for 6 3/4 diameter curve intersects the line for 1 hp.

The answer is (B).

187. The scaling laws and performance curves for pumps are given in the *NCEES Handbook*, Fluid Mechanics section. The power requirement of the new pump can be determined using the relation

$$\dot{W}_2 = \dot{W}_1\left(\frac{\rho_2}{\rho_1}\right)\left(\frac{N_2}{N_1}\right)^3\left(\frac{D_2}{D_1}\right)^5$$

step 1: Since both pumps are used to pump water, the fluid density is

$$\rho_2 = \rho_1 = 1$$

step 2: Apply the given pump properties to determine the power requirement.

$$\begin{aligned}\dot{W}_2 &= (1\ \text{hp})\left(\frac{1}{1}\right)\left(\frac{2000}{1500}\right)^3\left(\frac{8\ \text{in}}{6\ \text{in}}\right)^5 \\ &= (1\ \text{hp})(1.333)^3(1.333)^5 \\ &= (1.333)^8 \\ &= 9.969\ \text{hp}\quad (10\ \text{hp})\end{aligned}$$

The answer is (D).

188. The basis of dimensional analysis is that all terms of any correct physical equation must be dimensionally homogeneous. This implies that the power to which fundamental dimensions are raised must be the same on both the prototype and its model. Since gravity is the dominant force in the design of dams, the Froude number is be used for similitude. The Froude number is defined as the ratio of inertial force to gravity force.

step 1: Use Froude number formulas given in the *NCEES Handbook*, Fluid Mechanics section.

$$[\text{Fr}]_p = [\text{Fr}]_m$$

It is expressed as

$$\left[\frac{\text{v}^2}{lg}\right]_p = \left[\frac{\text{v}^2}{lg}\right]_m$$

$$\frac{\text{v}_p}{\text{v}_m} = \sqrt{\frac{l_p}{l_m}} = \sqrt{\frac{16}{1}} = 4$$

step 2: Using the above equation, determine flow velocity for model.

$$\begin{aligned}\text{v}_m &= \frac{\text{v}_p}{4} \\ &= \frac{4\ \text{fps}}{4} \\ &= 1\ \text{ft/sec}\end{aligned}$$

The answer is (B).

189.

step 1: See the *NCEES Handbook*, Thermodynamics and the Units and Conversion Factors sections. To determine the mass, m, use the ideal gas relation.

$$\begin{aligned}PV &= mRT \\ m &= \frac{PV}{RT}\end{aligned}$$

step 2: It is given that the initial pressure is

$$\begin{aligned}P &= 750\ \text{mm of Hg} \\ &= (750\ \text{mm of Hg})\left(\frac{1\ \text{atm}}{76.0\ \text{cm of Hg}}\right) \\ &\quad \times\left(\frac{1\ \text{cm}}{10\ \text{mm}}\right)\left(1.013\times 10^5\ \frac{\text{Pa}}{\text{atm}}\right) \\ &= 9.997\times 10^4\ \text{Pa}\quad (9.997\times 10^4\ \text{N/m}^2)\end{aligned}$$

step 3: Convert the temperature to kelvins.

$$\begin{aligned}T &= 19^\circ\text{C} \\ &= 19^\circ\text{C} + 273 = 292\text{K}\end{aligned}$$

step 4: The vessel volume is given in the Mensuration of Areas and Volumes table in the *NCEES Handbook*. For a sphere,

$$\begin{aligned}V &= \frac{\pi d^3}{6} \\ &= \frac{\pi(2\ \text{m})^3}{6} \\ &= 4.189\ \text{m}^3\end{aligned}$$

step 5: The molecular weight of air is 28.966 g/mol, as given in the Environmental Engineering section of the *NCEES Handbook*. The universal gas constant, R, is given by

$$
\begin{aligned}
R &= \frac{\text{Universal Gas Constant } (\overline{R})}{\text{Molecular weight of air}} \\
&= \frac{8314\ \dfrac{\text{J}}{\text{kmol·K}}}{28.966\ \dfrac{\text{g}}{\text{mol}}} \\
&= 287.0\ \frac{\text{kJ}}{\text{kg·K}} \\
&= 287.0\ \text{N·m/kg·K}
\end{aligned}
$$

step 6: Use the parameters computed above to determine mass.

$$
\begin{aligned}
m &= \frac{\left(9.997 \times 10^4\ \dfrac{\text{N}}{\text{m}^2}\right)(4.189\ \text{m}^3)}{\left(287.0\ \dfrac{\text{N·m}}{\text{kg·K}}\right)(292\text{K})} \\
&= 4.997\ \text{kg} \quad (5.0\ \text{kg})
\end{aligned}
$$

The answer is (D).

190. See the *NCEES Handbook*, Thermodynamics section. The problem states a special case of closed systems (with no change in kinetic or potential energy). The pressure is constant and energy is used to compress the gas. The work done is the hatched area shown.

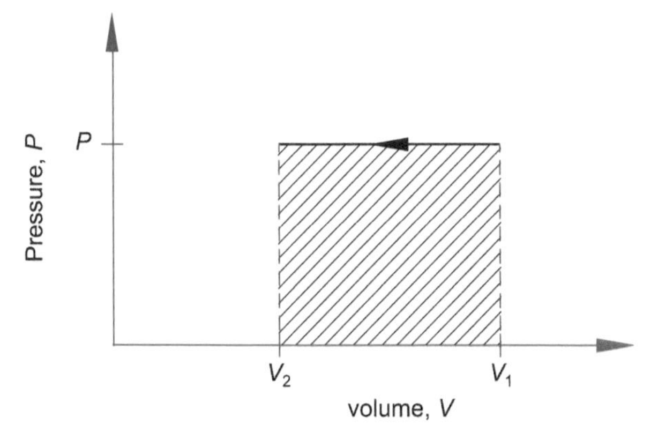

$$
\begin{aligned}
\text{Work done} &= \text{Pressure} \times \text{change in volume} \\
&= P(V_1 - V_2)
\end{aligned}
$$

step 1: Determine pressure in Pascal units.

$$
\begin{aligned}
P &= 2\ \text{bar} \\
&= (2\ \text{bar})\left(\frac{1 \times 10^5\ \text{Pa}}{1\ \text{bar}}\right) \\
&= 2 \times 10^5\ \text{Pa}
\end{aligned}
$$

step 2: Determine work done in kilojoules.

$$
\begin{aligned}
\text{Work done} &= P(V_1 - V_2) \\
&= (2 \times 10^5\ \text{Pa})(1\ \text{m}^3 - 0.8\ \text{m}^3) \\
&= 40{,}000\ \text{J} \\
&= 40\ \text{kJ}
\end{aligned}
$$

The answer is (D).

191. An isentropic process is a process in which the working substance neither receives heat nor gives heat to its surrounding. The process takes place without a change of entropy. It is a special case of an adiabatic process that is reversible. The work transfers of the system are frictionless, and there is no transfer of heat or matter. As such, it is an idealized process. The problem states a special case of closed systems (with no change in kinetic or potential energy). The temperature is constant, and energy is used to compress the gas. The work done is the hatched area shown. See the *NCEES Handbook*, Thermodynamics section.

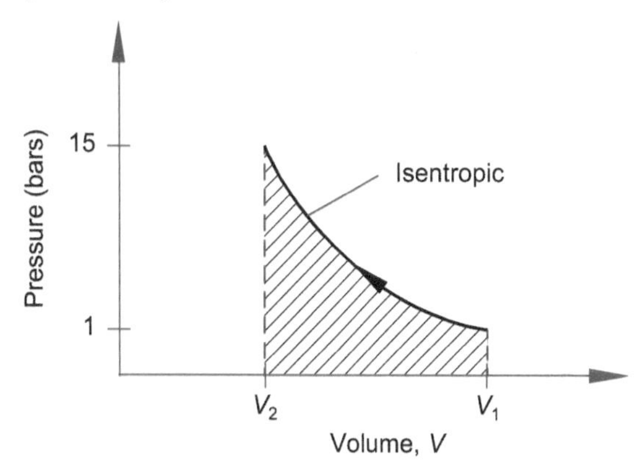

step 1: Determine volume in the compressed state. It is given that for the isentropic process

$$pV^k = \text{constant}$$

$$k = \frac{c_p}{c_v}$$

Use the Thermal and Physical Property Tables given in the *NCEES Handbook*. For propane, $k = 1.12$.

It is given that the process is isentropic. Therefore,

$$p_1V_1^k = p_2V_2^k$$
$$\left(\frac{V_1}{V_2}\right)^k = \frac{p_2}{p_1} = \frac{10 \text{ bar}}{1 \text{ bar}} = 10$$
$$\left(\frac{V_1}{V_2}\right) = 10^{1/k} = 10^{1/1.12} = 7.814$$

The initial volume is known. Calculate the final volume.

$$V_2 = \frac{V_1}{7.814} = \frac{1 \text{ m}^3}{7.814} = 0.1280 \text{ m}^3$$

step 2: Determine the final temperature using Charles' law.

$$\frac{p_1V_1}{T_1} = \frac{p_2V_2}{T_2}$$
$$T_2 = T_1\left(\frac{p_2V_2}{p_1V_1}\right)$$

All parameters are known. Therefore, the final temperature is

$$T_2 = (289\text{K})\frac{(10 \text{ bar})(0.1280 \text{ m}^3)}{(1 \text{ bar})(1 \text{ m}^3)} = 369.9\text{K} \quad (370\text{K})$$

step 3: Determine change in temperature.

$$\Delta T = T_2 - T_1 = 370\text{K} - 289\text{K} = 81°\text{C}$$

The answer is (D).

192. The problem states a steady flow system which does not change with time. See the *NCEES Handbook*, Thermodynamics section. The system general equation is given by

$$\sum \dot{m}_i\left(h_i + \frac{\text{v}_i^2}{2} + gZ_i\right) - \sum \dot{m}_e\left(h_e + \frac{\text{v}_e^2}{2} + gZ_e\right) + \dot{Q}_{\text{in}} - \dot{W}_{\text{out}} = 0$$

step 1: In this case, the mass at the inlet equals the mass at the exit, and there is no change in its elevation. Therefore, there is no change in the potential energy. Simplify the equation to determine work done.

$$\dot{W}_{\text{out}} = \dot{m}\left(h_i - h_e + \frac{\text{v}_i^2 - \text{v}_e^2}{2}\right) + \dot{Q}_{\text{in}}$$

step 2: Determine the net rate of work out of the system.

$$\begin{aligned} \text{Enthalpy, } h_i &= 1000 \text{ Btu/lbm} \\ h_e &= 800 \text{ Btu/lbm} \\ \text{Steam velocity, } \text{v}_i &= 50 \text{ ft/sec} \\ \text{v}_e &= 100 \text{ ft/sec} \\ \text{Heat loss, } \dot{Q} &= 49 \text{ Btu/sec} \end{aligned}$$

Use the conversion factor from the *NCEES Handbook* to convert ft-lbf units to Btu units. Apply the formula.

$$\dot{W}_{\text{out}} = \left(5 \frac{\text{lbm}}{\text{sec}}\right) \times \left(1000 \frac{\text{Btu}}{\text{lbm}} - 800 \frac{\text{Btu}}{\text{lbm}} + \frac{\left(50 \frac{\text{ft}}{\text{sec}}\right)^2 - \left(100 \frac{\text{ft}}{\text{sec}}\right)^2}{(2)\left(32.2 \frac{\text{lbm-ft}}{\text{lbf-sec}^2}\right) \times \left(778 \frac{\text{ft-lbf}}{\text{Btu}}\right)}\right) - 49 \frac{\text{Btu}}{\text{sec}}$$
$$= 950.3 \text{ Btu/sec} \quad (950 \text{ Btu/sec})$$

The answer is (B).

193. The problem involves reading values off the steam tables or the Mollier diagram. See the *NCEES Handbook*, Thermodynamics section, steam tables, Mollier (h, s) diagram for steam, and Rankine cycle diagrams. Turbines are considered adiabatic systems. There is no heat transfer in a turbine, and the entropy change is considered zero. Changes in enthalpies are considered to determine a turbine's isentropic efficiency.

$$\eta = \frac{h_i - h_e}{h_i - h_{es}}$$

Steam tables provide the needed values to calculate η. However, the tables given in the *NCEES Handbook* are in SI units, and the problem is stated in Btu units. The unit conversion is time consuming. Use the Mollier diagram that is in Btu units.

step 1: Using the Mollier diagram, locate the point at the intersection of the saturation line and 200 psi pressure line and determine the corresponding enthalpy.

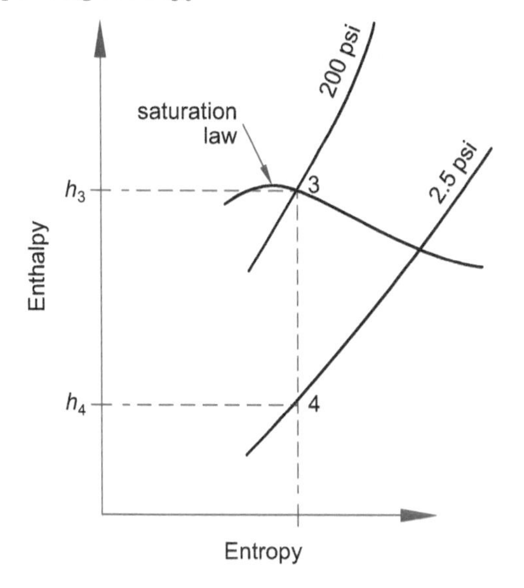

$$h_i = h_3 = 1200 \text{ Btu/lbm}$$

step 2: Keeping the entropy constant as determined in step 1, draw a straight line down to the 2.5 psi pressure line, noted as point 4 as shown. The corresponding enthalpy is

$$h_e = h_4 = 910 \text{ Btu/lbm}$$

step 3: Use the exit saturated liquid enthalpy level given the problem statement, $h_{4s} = 102$ Btu/lbm. Use the enthalpy values in Btu/lbm and determine the turbine efficiency.

$$\begin{aligned}\eta &= \frac{1200 \text{ Btu/lbm} - 910 \text{ Btu/lbm}}{1200 \text{ Btu/lbm} - 102 \text{ Btu/lbm}} \\ &= 0.2641 \quad (26\%)\end{aligned}$$

The answer is (A).

194. A diesel cycle is shown.

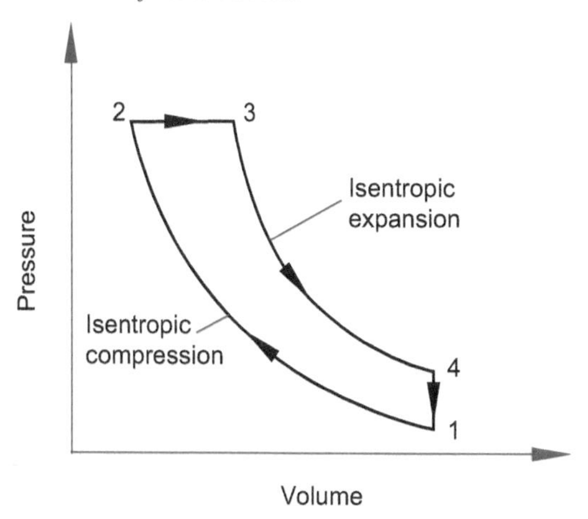

An ideal diesel cycle consists of two reversible adiabatic or isentropic processes, a constant-pressure process, and a constant-volume process. See the *NCEES Handbook*, Mechanical Engineering section. The engine efficiency is

$$\eta = 1 - \left(\frac{1}{r^{k-1}}\right)\left(\frac{r_c^k - 1}{k(r_c - 1)}\right)$$

step 1: From Thermal and Physical Property Tables (at room temperature), the ratio k for air = 1.4.

step 2: It is given that the volume ratio is

$$r = \frac{V_1}{V_2} = \frac{10}{1} = 10$$

step 3: Determine, the cutoff ratio, r_c. Let $V_2 = 1$ unit. Therefore, $V_1 = 10$ units. The volume at fuel cutoff is

$$\begin{aligned}V_3 &= V_2 + 0.07(V_1 - V_2) \\ &= 1 + 0.07(10 - 1) \\ &= 1.63\end{aligned}$$

The cutoff ratio is

$$r_c = \frac{V_3}{V_2} = \frac{1.63}{1} = 1.63$$

The efficiency is

$$\begin{aligned}\eta &= 1 - \left(\frac{1}{(10)^{1.4-1}}\right)\left(\frac{(1.63)^{1.4} - 1}{(1.4)(1.63 - 1)}\right) \\ &= 1 - 0.4432 \\ &= 0.5568 \quad (56\%)\end{aligned}$$

The answer is (C).

195. The P-h Diagram for Refrigerant HFC-134a is given in the *NCEES Handbook*, Thermodynamics section. Refrigeration cycles are the reverse of heat engine cycles. Heat is moved from low to high temperature, requiring work, W. The coefficient of performance is defined as

$$\text{COP} = \frac{h_1 - h_4}{h_2 - h_1}$$

The determination of COP requires enthalpies (h-values) at four corners of the refrigeration cycle. Use the p-h diagram for HFC-134a.

step 1: For 0.1 MPa pressure, locate the intersection of the pressure line and the saturated vapor side of the saturation dome.

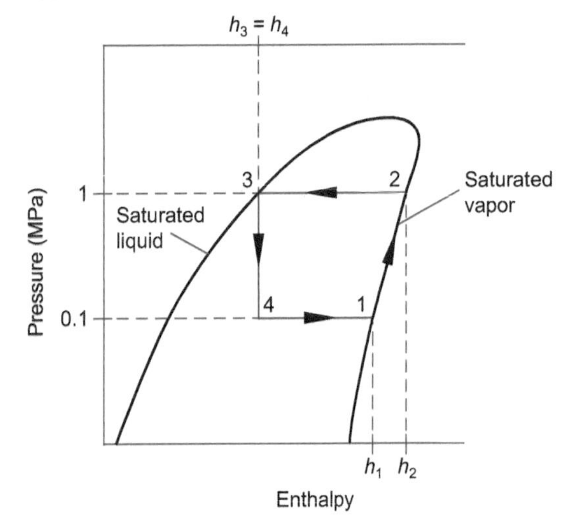

Read the enthalpy value on the x-axis. It is

$$h_1 = 380 \text{ kJ/kg}$$

step 2: For 1 MPa pressure, locate the intersection of the line and the saturated vapor side of the saturation dome. Read the enthalpy value on the x-axis. It is

$$h_2 = 430 \text{ kJ/kg}$$

step 3: For 1 MPa pressure line, locate the intersection of the line and the saturated liquid side of the saturation dome. Read the enthalpy value on the x-axis.

$$h_3 = 255 \text{ kJ/kg}$$

step 4: By definition, enthalpies h_3 and h_4 are equal.

$$h_4 = 255 \text{ kJ/kg}$$

step 5: Apply the formula to calculate COP.

$$\begin{aligned}\text{COP} &= \frac{h_1 - h_4}{h_2 - h_1} \\ &= \frac{380 \text{ kJ/kg} - 255 \text{ kJ/kg}}{430 \text{ kJ/kg} - 380 \text{ kJ/kg}} = 2.5\end{aligned}$$

The answer is (D).

196. See the *NCEES Handbook*, Thermodynamics section. It is given that the ice forms at 0°C, and

$$\text{Specific heat of water, } c_p = 4.18 \text{ kJ/kg}$$

step 1: Convert daily production to production per second.

$$\begin{aligned}\dot{m} &= \frac{10{,}000 \ \frac{\text{kg}}{\text{d}}}{\left(24 \ \frac{\text{h}}{\text{d}}\right)\left(60 \ \frac{\text{min}}{\text{h}}\right)\left(60 \ \frac{\text{s}}{\text{min}}\right)} \\ &= 0.1157 \text{ kg/s}\end{aligned}$$

step 2: Two types of heat need to be extracted from the incoming water.

$$\dot{Q} = \dot{m}\left(\begin{array}{r}\text{Specific heat to drop water} \\ \text{temp from 23.2°C to 0°C} \\ + \text{ Latent heat to convert} \\ \text{liquid to solid}\end{array}\right)$$

Apply the above values and calculate.

$$\begin{aligned}\dot{Q} &= \left(0.1157 \ \frac{\text{kg}}{\text{s}}\right) \\ &\quad \times \left(\begin{array}{c}\left(4.18 \ \frac{\text{kJ}}{\text{kg·°C}}\right)(23.2\text{°C} - 0\text{°C}) \\ + \ 335 \ \frac{\text{kJ}}{\text{kg}}\end{array}\right) \\ &= \left(0.1157 \ \frac{\text{kg}}{\text{s}}\right)\left(432.0 \ \frac{\text{kJ}}{\text{kg}}\right) \\ &= 49.98 \text{ kJ/s} \quad (50 \text{ kW})\end{aligned}$$

The answer is (A).

197. See the *NCEES Handbook*, Thermodynamics section. Use the psychrometric chart and read off the results as shown. The chart is a plot of atmospheric air properties as a function of dry-bulb temperature (t_{db}).

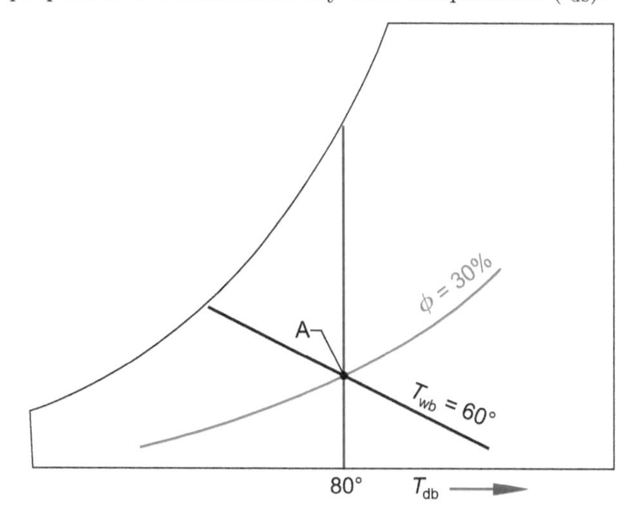

step 1: The air condition is 80°F dry-bulb temperature, T_{db}, and 60° F wet-bulb temperature, T_{wb}. Locate the condition on the chart at point A.

step 2: Read the relative humidity along the relative humidity curve. It is 30%.

The answer is 30%.

198. See the *NCEES Handbook*, Chemistry and Biology and Thermodynamics sections. The stoichiometric equation for combustion of carbon in air given in the *Handbook* is

$$C+O_2 \rightarrow CO_2$$

The equation shows that it takes two volumes of oxygen to burn one volume of carbon. Carbon monoxide is produced instead of carbon dioxide if the oxygen supply is insufficient. Therefore, determine whether the 4 kg of oxygen is sufficient to complete the combustion. To calculate mass products of combustion, use the periodic table of elements given in the Chemistry section of the *NCEES Handbook*. In terms of moles, the above equation becomes

$$\begin{aligned} 1 \text{ mol C} + 1 \text{ mol } O_2 &= 1 \text{ mol } CO_2 \\ 12 \text{ kg} + 32 \text{ kg} &= 44 \text{ kg} \end{aligned}$$

To determine the products of combustion for 1 kg carbon, divide both sides of the equation by 12.

$$\frac{12 \text{ kg}}{12} + \frac{32 \text{ kg}}{12} = \frac{44 \text{ kg}}{12}$$

$$1 \text{ kg} + \frac{8}{3} \text{ kg} = \frac{11}{3} \text{ kg} < 4 \text{ kg}$$

Compare the supply of oxygen with the amount required for complete combustion. Since the oxygen supply is more than that needed for complete combustion, no carbon monoxide is produced during the combustion process.

The answer is 0 kg.

199. See the *NCEES Handbook*, Heat Transfer section, for conduction heat loss equations. For a plane wall

$$\text{Thermal resistance, } R = \frac{\text{Thickness}}{\text{conductivity}} = \frac{L}{kA}$$

It is measured in either K/W or °C/W, as the interval difference in degrees Celsius or kelvins in conduction is the same. A composite wall offers conduction in series as shown.

$\dot{Q}$ → $T_{\infty 1}$ — $\frac{L_1}{k_1 A}$ — $\frac{L_2}{k_2 A}$ — $\frac{L_3}{k_3 A}$ — $T_{\infty 2}$

For a composite wall, thermal resistance is

$$\sum R = \sum \frac{L}{kA}$$

For unit wall area,

$$\begin{aligned} \sum \frac{L}{k} &= \frac{L_1}{k_1} + \frac{L_2}{k_2} + \frac{L_3}{k_3} \\ &= \frac{0.1 \text{ m}}{0.1 \text{ W/m·K}} + \frac{0.035 \text{ m}}{0.07 \text{ W/m·K}} + \frac{0.025 \text{ m}}{0.05 \text{ W/m·K}} \\ &= 2 \text{ K/W} \quad (2 \text{ °C/W}) \end{aligned}$$

The answer is (B).

200. See the *NCEES Handbook*, Heat Transfer section, for two applicable formulas. The shape factor, view factor, or configuration factor is the fraction of the radiation leaving surface 1 that is intercepted by the other surfaces. Its value is between 0 and 1. In an enclosure, the conservation rule applies. It states that all radiation leaving surface i must be intercepted by the enclosed surfaces. The summation rule for n surfaces is

$$\sum_{j=1}^{n} F_{ij} = 1$$

step 1: In this case, the enclosure has three surfaces. For radiation leaving surface 1,

$$\sum_{j=1}^{j=3} F_{ij} = F_{11} + F_{12} + F_{13} = 1$$

The shape factor F_{11} represents the fraction of radiation that leaves surface 1 and then is intercepted by the same surface. Because surface 1 is flat,

$$F_{11} = 0$$

By symmetry,

$$F_{12} = F_{13}$$

Therefore,

$$\sum_{j=1}^{j=3} F_{ij} = 2F_{12} = 1$$

$$F_{12} = 0.5$$

step 2: To determine shape factor F_{21}, consider the reciprocity relationship. For two surfaces that are diffuse emitters and reflectors and have uniform radiosity, the reciprocity relationship applies and is given by

$$A_i F_{ij} = A_j F_{ji}$$

Consider a unit length of the air duct and determine the areas of surfaces 1 and 2.

$$A_2 = L \times 1 = L$$
$$A_1 = \sqrt{2}\,L \times 1 = \sqrt{2}\,L$$

Calculate shape factor F_{21}.

$$A_1 F_{12} = A_2 F_{21}$$
$$(\sqrt{2}\,L)(0.5) = LF_{21}$$
$$F_{21} = \frac{(\sqrt{2}\,L)(0.5)}{L} = 0.7071 \quad (0.71)$$

The answer is (C).

201. Use the *NCEES Handbook*, Heat Transfer section. The net energy exchange by radiation between two diffuse-gray surfaces that form an enclosure is expressed as

$$\dot{Q}_{12} = \frac{\sigma(T_1^4 - T_2^4)}{\dfrac{1-\varepsilon_1}{A_1\varepsilon_1} + \dfrac{1}{A_1F_{12}} + \dfrac{1-\varepsilon_2}{\varepsilon_2 A_2}}$$

step 1: Consider the length of each tube, $L = 1$ m. Use m as the length unit. All parameters can be calculated form the data provided.

$$\begin{aligned}
\sigma &= \text{Stefan-Boltzmann constant} \\
&= 5.67 \times 10^{-8}\ \text{W/m}^2\text{·K}^4 \\
d_1 &= (20\ \text{mm})\left(\frac{1\ \text{m}}{1000\ \text{mm}}\right) = 0.02\ \text{m} \\
d_2 &= (40\ \text{mm})\left(\frac{1\ \text{m}}{1000\ \text{mm}}\right) = 0.04\ \text{m} \\
\text{Surface Area } A_1 &= \pi d_1 L = \pi(0.02\ \text{m})(1\ \text{m}) \\
&= 0.02\pi\ \text{m}^2 \\
\text{Surface Area } A_2 &= \pi d_2 L = \pi(0.04\ \text{m})(1\ \text{m}) \\
&= 0.04\pi\ \text{m}^2 \\
T_1 &= 100\text{K} \\
T_2 &= 300\text{K} \\
\varepsilon &= \text{emissivity of all surfaces} = 0.2
\end{aligned}$$

step 2: The shape factor F_{12} is the fraction of the radiation leaving surface 1 that is intercepted by surface 2. From the conservation equation, its value is 1.

step 3: Assume that the radiation energy is flowing from surface 1 to surface 2—from the outer surface of the inner tube to the inner surface of the outer tube. Determine the heat gain using the energy exchange formula.

$$\dot{Q}_{12} = \frac{\sigma(T_1^4 - T_2^4)}{\dfrac{1-\varepsilon_1}{A_1\varepsilon_1} + \dfrac{1}{A_1F_{12}} + \dfrac{1-\varepsilon_2}{\varepsilon_2 A_2}}$$

$$= \frac{\left(\begin{array}{c}(5.67 \times 10^{-8}\ \text{W/m}^2\text{·K}^4) \\ \times\left((100\text{K})^4 - (300\text{K})^4\right)\end{array}\right)}{\left(\begin{array}{c}\dfrac{1-0.2}{(0.02\pi\ \text{m}^2)(0.2)} + \dfrac{1}{(0.02\pi\ \text{m}^2)(1.0)} \\ + \dfrac{1-0.2}{(0.2)(0.04\pi\ \text{m}^2)}\end{array}\right)}$$

$$= \pi\left(\frac{(5.67\ \text{W/m}^2)(-80)}{200\ \text{m}^2 + 50\ \text{m}^2 + 100\ \text{m}^2}\right)$$

$$= -4.072\ \text{W per meter of tube length} \quad (-4.1\ \text{W/m})$$

The negative sign indicates that the radiation energy is traveling from surface 2 to surface 1 and not from surface 1 to surface 2.

The answer is (B).

202. See the *NCEES Handbook,* Heat Transfer section, for convection and radiation heat loss equations.

$$\begin{aligned}\text{Total heat loss, } \dot{Q} &= \text{Convection loss} \\ &\quad + \text{Radiation loss} \\ &= hA(T_w - T_\infty) \\ &\quad + \varepsilon\sigma A(T_1^{\ 4} - T_2^4)\end{aligned}$$

The heat loss through convection and radiation are calculated using the absolute temperature scale (kelvins). For convection, the heat loss is based on the linear difference of the absolute temperatures. The temperature difference is numerically the same for both the Celsius and Kelvin scales. To avoid confusion in units, use kelvins for both convection and radiation losses.

See the *NCEES Handbook*, the Units and Conversion Factors: Fundamental Constants section, for the Stefan-Boltzmann constant.

$$\sigma = 5.67 \times 10^{-8}\ \text{W}/(\text{m}^2\cdot\text{K}^4)$$

It is given that

$$\text{Emissivity for steel pipe outer surface}, \varepsilon = 0.6$$

step 1: Consider a 1 m long pipe section.

$$\begin{aligned}\text{Surface Area} &= \pi DL \\ &= \pi\left((500\ \text{mm})\left(\frac{1\ \text{m}}{1000\ \text{mm}}\right)\right)(1\ \text{m}) \\ &= 1.571\ \text{m}^2\end{aligned}$$

Convection coefficient,

$$h = 15\ \text{W}/\text{m}^2\cdot\text{K}$$

Pipe surface temp,

$$\begin{aligned}T_w &= 100°\text{C} \\ &= 100°\text{C} + 273 \\ &= 373\text{K}\end{aligned}$$

Room and wall temp,

$$\begin{aligned}T_\infty &= 27°\text{C} \\ &= 27°\text{C} + 273 \\ &= 300\text{K}\end{aligned}$$

step 2: All parameters are known. Determine the heat loss.

$$\begin{aligned}\dot{Q} &= hA(T_w - T_\infty) + \varepsilon\sigma A(T_1^{\ 4} - T_2^4) \\ &= \left(\begin{array}{l}(15\ \text{W}/\text{m}^2\cdot\text{K})(1.571\ \text{m}^2) \\ \quad \times(373\text{K} - 300\text{K}) \\ \quad + (0.6)\left(5.67\times 10^{-8}\ \dfrac{\text{W}}{\text{m}^2\cdot\text{K}^4}\right) \\ \quad \times(1.571\ \text{m}^2)\left(\begin{array}{l}(373\text{K})^4 \\ \quad -(300\text{K})^4\end{array}\right)\end{array}\right) \\ &= 1720.25\ \text{W}/\text{m} + 601.63\ \text{W}/\text{m} \\ &= 2322\ \text{W}/\text{m} \quad (2320\ \text{W}/\text{m})\end{aligned}$$

The answer is (D).

203. The pertinent properties of iron are given in the *NCEES Handbook*, Properties of Metals table.

Density,

$$\rho = 7873\ \text{kg}/\text{m}^3$$

Specific heat,

$$c = 456.4\ \text{J}/\text{kg}\cdot\text{K}$$

Heat conductivity,

$$k = 83.5\ \text{W}/\text{m}\cdot\text{K}$$

step 1: The Fourier number is a dimensionless number. It is defined as

$$\text{Fo} = \frac{\text{Heat conduction rate}}{\text{Thermal Energy storage in a solid}}$$

It is given in the *NCEES Handbook*, Heat Transfer: Approximate Solution for Solid with Sudden Convection section as

Fourier number,

$$\text{Fo} = \frac{\alpha t}{L^2}$$

Thermal diffusivity,

$$\alpha = \frac{k}{\rho c}$$

step 2: With all the parameters known, determine the value of Fo.

$$t = 5.5 = (5.5\ \text{min})\left(60\ \frac{\text{s}}{\text{min}}\right) = 330\ \text{s}$$

$$L = (35\ \text{mm})\left(\frac{1\ \text{m}}{1000\ \text{mm}}\right) = 0.035\ \text{m}$$

$$\text{Fo} = \frac{83.5\ \text{W/m·K}}{(7873\ \text{kg/m}^3)(456.4\ \text{J/kg·K})}\left(\frac{330\ \text{s}}{(0.035\ \text{m})^2}\right) = 6.260 \quad (6.3)$$

The answer is (B).

204. Heat transfer is affected by the specific heat values of the fluids in the exchanger. The specific heat capacity of the oil is given in the problem statement. For the specific heat capacity of water, see the *NCEES Handbook*, Thermodynamics section.

For water,

$$c_{p,\text{water}} = 4.18\ \text{kJ/kg·K}$$
$$c_{p,\text{oil}} = 2\ \text{kJ/kg·K}$$

step 1: The rate of heat transfer associated with either fluid flow stream in a heat exchanger is

$$\dot{Q} = \dot{m}c_p(T_{\text{exit}} - T_{\text{inlet}})$$

Based on the assumption described, the heat balance of both fluids is expressed as

$$\dot{m}_{\text{oil}}c_{p,\text{oil}}(T_{H,i} - T_{H,o}) = \dot{m}_{\text{water}}c_{p,\text{water}}(T_{c,o} - T_{c,i})$$

The flow rates of water and oils are given. Out of the four temperatures noted in the above equation, three temperatures are given, and only $T_{c,i}$ is unknown.

step 2: Find the temperature, $T_{c,i}$. Since the losses are negligible, use the conservation of energy equation, and find the water temperature at the inlet. Rearrange the equation to solve for $T_{c,i}$.

$$T_{c,i} = T_{c,o} - \frac{\dot{m}_{\text{oil}}c_{p,\text{oil}}(T_{H,i} - T_{H,o})}{\dot{m}_{\text{water}}c_{p,\text{water}}}$$
$$= 40°\text{C} - \frac{\dot{m}_{\text{oil}}c_{p,\text{oil}}(100°\text{C} - 54°\text{C})}{\dot{m}_{\text{water}}c_{p,\text{water}}}$$
$$= 40°\text{C} - \left(\left(\frac{1\ \frac{\text{kg}}{\text{s}}}{2\ \frac{\text{kg}}{\text{s}}}\right)\left(\frac{2\ \text{kJ/kg·K}}{4.18\ \text{kJ/kg·K}}\right) \times (100°\text{C} - 54°\text{C})\right)$$
$$= 40°\text{C} - 11.00°\text{C}$$
$$= 29°\text{C}$$

The answer is (C).

205. See the *NCEES Handbook*, Heat Transfer section. The Biot number is used in transient conduction problems in which a solid body experiences a sudden change in its thermal environment. It is defined as

$$\text{Bi} = \frac{hV}{kA_s}$$

See the *NCEES Handbook*, Mathematics section. For a sphere,

$$V = \text{Volume of a sphere} = \frac{\pi d^3}{6}$$
$$A_s = \text{Surface area of a sphere} = \pi d^2$$
$$\frac{V}{A_s} = \frac{\frac{\pi d^3}{6}}{\pi d^2} = \frac{d}{6} = \frac{9\ \text{mm}}{6} = 1.5\ \text{mm} = 1.5 \times 10^{-3}\ \text{m}$$

It is given that

$$h = \text{Convection heat-transfer coefficient of the gas} = 400\ \text{W/m}^2\text{·K}$$
$$k = 20\ \text{W/m·K}$$

Therefore,

$$\begin{aligned}\mathrm{Bi} &= \frac{hV}{kA_s}\\ &= \frac{(400\ \mathrm{W/m^2{\cdot}K})(1.5\times 10^{-3}\ \mathrm{m})}{20\ \mathrm{W/m{\cdot}K}}\\ &= 0.03\end{aligned}$$

The answer is 0.03.

206. See the *NCEES Handbook*, Instrumentation, Measurement, and Control section.

step 1: Select a subsystem out of the given system block diagram for which the overall response is known. Refer to the *NCEES Handbook*, Instrumentation, Measurement, and Control section. Use the formula to determine the overall function. The response for a subsystem with transfer functions G_1 and G_2 is well known and is given in the *NCEES Handbook*. Select subsystem A as shown.

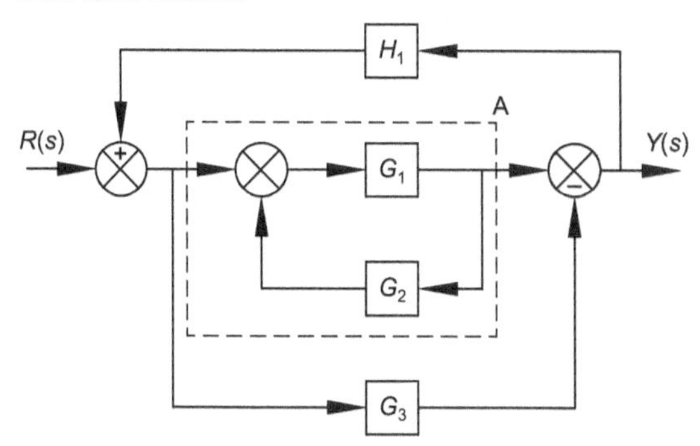

For a system that has a controller G_1 and a negative feedback G_2, the overall transfer function is given by

$$Y_A(s) = \frac{G_1}{1+G_1G_2}$$

step 2: Show the system with results from step 1 incorporated. Select next subsystem, B. It sums up two systems, Y_A and G_3, that are in parallel. Their resultant is their algebraic sum.

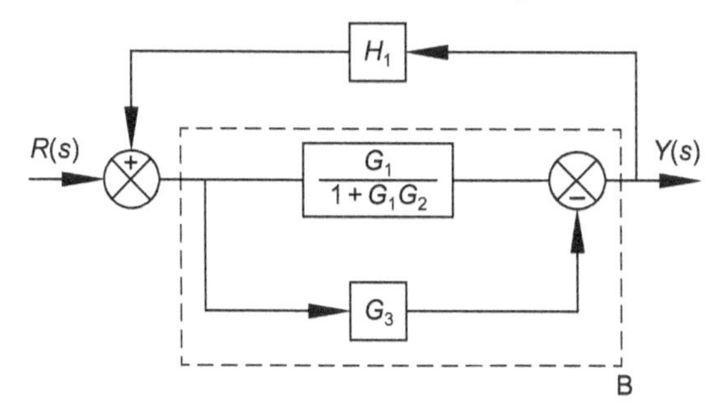

step 3: Use the formula and substitute transfer function values to determine overall function of the subsystem B block output.

$$\begin{aligned}Y_B(s) &= \frac{G_1}{1+G_1G_2} + G_3\\ &= \frac{G_1 + G_1G_2G_3}{1+G_1G_2}\end{aligned}$$

step 4: Now the system is reduced to the two subsystems: Y_B and the feedback H_1. The feedback is positive and, therefore, the sign in the denominator in the overall response is negative.

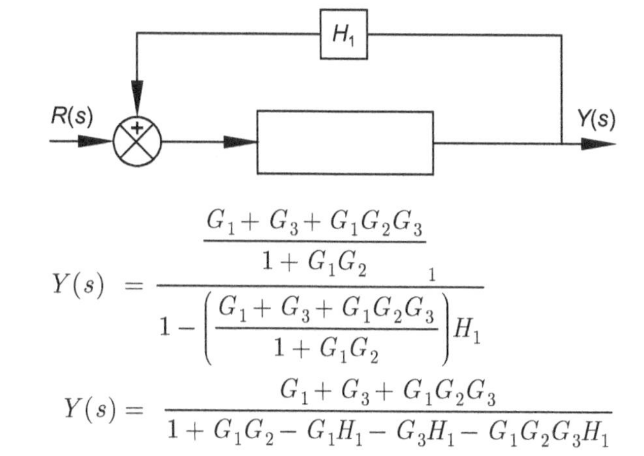

$$Y(s) = \frac{\dfrac{G_1+G_3+G_1G_2G_3}{1+G_1G_2}}{1-\left(\dfrac{G_1+G_3+G_1G_2G_3}{1+G_1G_2}\right)H_1}$$

$$Y(s) = \frac{G_1+G_3+G_1G_2G_3}{1+G_1G_2-G_1H_1-G_3H_1-G_1G_2G_3H_1}$$

The answer is

$$A = G_1H_1$$
$$B = G_3H_1$$

207. See the *NCEES Handbook*, Instrumentation, Measurement, and Control section. For a system that has a controller $G_1(s)$ and a negative feedback $H(s)$, the overall transfer function is given by

$$Y(s) = \frac{G_1(s)}{1+G_1(s)H(s)}$$

step 1: Use the formula to determine overall function. Since $H(s)$ is not specified, it is assumed as 1.

$$Y(s) = \frac{\dfrac{18}{s+2}}{1+\left(\dfrac{18}{s+2}\right)(1)}$$

To simply the equation, multiply both the numerator and denominator by $s+2$.

$$Y(s) = \frac{(s+2)\left(\frac{18}{s+2}\right)}{(s+2)\left(1+\left(\frac{18}{s+2}\right)\right)} = \frac{18}{s+20}$$

step 2: For steady state, $s = 0$.

$$Y(s) = \frac{18}{0+20} = 0.9$$

The answer is (A).

208. See the *NCEES Handbook*, Instrumentation, Measurement, and Control section. Measurement accuracy is defined as closeness of agreement between a measured quantity value and a true quantity value of a measurement. When reporting measurement results, it is necessary to provide an associated uncertainty so that those who use it may assess its reliability. One method to assess the uncertainty is to use the Kline-McClintock general equation.

$$w_R = \sqrt{\left(w_1\frac{\partial f}{\partial x_1}\right)^2 + \left(w_2\frac{\partial f}{\partial x_2}\right)^2 + \ldots + \left(w_n\frac{\partial f}{\partial x_n}\right)^2}$$

In this case, uncertainties in mass and velocity measurements are known. The uncertainty in measurement of kinetic energy (KE) needs to be computed.

step 1: Determine the average KE of the object.

$$\begin{aligned} \text{KE} &= \frac{1}{2}m\text{v}^2 \\ &= \frac{1}{2}(22.5\ \text{kg})\left(5.19\ \frac{\text{m}}{\text{s}}\right)^2 \\ &= 303.0\ \text{J} \quad (303\ \text{J}) \end{aligned}$$

step 2: See the *NCEES Handbook*, Mathematics section, for derivatives and partially differentiate the KE equation with respect to the mass and velocity.

$$\begin{aligned} \frac{\partial\text{KE}}{\partial m} &= \frac{1}{2}(1)\text{v}^2 = \frac{\text{v}^2}{2} \\ \frac{\partial\text{KE}}{\partial \text{v}} &= \frac{1}{2}m(2\text{v}) = m\text{v} \end{aligned}$$

step 3: Apply the Kline-McClintock equation to determine the measurement uncertainty in KE.

$$\begin{aligned} \delta\text{KE} &= \sqrt{\left(\frac{\partial\text{KE}}{\partial m}\,\delta m\right)^2 + \left(\frac{\partial\text{KE}}{\partial \text{v}}\,\delta\text{v}\right)^2} \\ &= \sqrt{\left(\frac{\text{v}^2}{2}\,\delta m\right)^2 + (m\text{v}\,\delta\text{v})^2} \\ &= \sqrt{\left(\left(\frac{\left(5.19\ \frac{\text{m}}{\text{s}}\right)^2}{2}\right)(0.1\ \text{kg})\right)^2 + \left((22.5\ \text{kg})\left(5.19\ \frac{\text{m}}{\text{s}}\right) \times\left(0.05\ \frac{\text{m}}{\text{s}}\right)\right)^2} \\ &= \sqrt{1.814\left(\frac{\text{kg}\cdot\text{m}^2}{\text{s}^2}\right)^2 + 34.09\left(\frac{\text{kg}\cdot\text{m}^2}{\text{s}^2}\right)^2} \\ &= \sqrt{35.904\left(\frac{\text{kg}\cdot\text{m}^2}{\text{s}^2}\right)^2} \\ &= \sqrt{35.904\ \text{J}^2} \\ &= 5.992\ \text{J} \quad (6\ \text{J}) \end{aligned}$$

step 4: The measured KE is 303 J ± 6 J.

The answer is 303 ± 6 J.

209. See the *NCEES Handbook*, Instrumentation, Measurement, and Control section. The gauge factor (GF) is defined as the ratio of fractional change in electrical resistance to the fractional change in length (strain).

$$\text{GF} = \frac{\frac{\Delta R}{R}}{\frac{\Delta L}{L}}$$

step 1: It is given that

$$\begin{aligned} R &= 350\ \Omega \\ \Delta R &= 0.28\ \Omega \\ L &= 10\ \text{in} \\ \text{GF} &= 2.0 \end{aligned}$$

All other parameters, except the bar elongation, are known.

step 2: Using the data, determine the elongation.

$$\begin{aligned}\Delta L &= \frac{L}{\text{GF}}\frac{\Delta R}{R}\\ &= \frac{10\text{ in}}{2}\left(\frac{0.28\ \Omega}{350\ \Omega}\right)\\ &= 0.004\text{ in}\\ &= 4000\ \mu\text{in}\end{aligned}$$

The answer is (C).

210. See the *NCEES Handbook*, Instrumentation, Measurement, and Control section. The Steinhart-Hart equation is the best mathematical expression for resistance temperature relationship of negative temperature coefficient (NTC) thermistors and NTC probe assemblies.

$$\frac{1}{T} = A + B\ln R + C(\ln R)^3$$

For a typical thermistor, the three constants, A, B, and C, in the equation are given in the *NCEES Handbook* as

$$\begin{aligned}A &= 1.403\times 10^{-3}\\ B &= 2.373\times 10^{-4}\\ C &= 9.827\times 10^{-8}\end{aligned}$$

step 1: For resistance $R = 9.8\text{ k}\Omega$, determine T.

$$\begin{aligned}\frac{1}{T} &= A + B\ln R + C(\ln R)^3\\ &= (1.403\times 10^{-3}) + (2.373\times 10^{-4})(\ln 9800)\\ &\quad +(9.827\times 10^{-8})(\ln 9800)^3\\ &= 0.003660\\ T &= \frac{1}{0.00366} = 273.22\text{K}\end{aligned}$$

step 2: The temperature is in kelvins. Convert kelvins to degrees Celsius.

$$\begin{aligned}T &= 273.22\text{ K} - 273.15\\ &= 0.07^\circ\text{C}\quad (0^\circ\text{C})\end{aligned}$$

The answer is 0°C.

211.

step 1: Draw a free-body diagram through section X-X, as shown. The section X-X is subjected to combined tension, F, and bending moment, M.

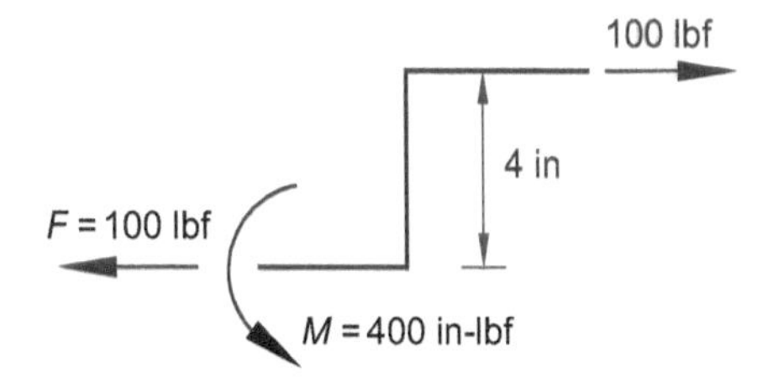

The von Mises stress is the maximum combined stress at the section, and is given by

$$\sigma' = \frac{P}{A} + \frac{Mc}{I}$$

The equation is given in the *NCEES Handbook*, Mechanical Engineering: Shafts and Axles section, in a general form.

$$\sigma' = \frac{4}{\pi d^3}\left((8M + Fd)^2 + (48T)^2\right)^{1/2}$$

In this problem, there is no torsion, T, on the section. Therefore, the *NCEES Handbook* equation can be simplified to

$$\sigma' = \frac{4}{\pi d^3}(8M + Fd)$$

step 2: All parameters are known. Therefore, the von Mises stress is

$$\begin{aligned}\sigma' &= \frac{4}{\pi(1\text{ in})^3}\left((8)(400\text{ in-lbf}) + (100\text{ lbf})(1\text{ in})\right)\\ &= \left(\frac{4}{\pi}(3300)\ \frac{\text{lbf}}{\text{in}^2}\right)\left(\frac{1\text{ kip}}{1000\text{ lbf}}\right)\\ &= 4.201\text{ kips/in}^2\quad (4.2\text{ ksi})\end{aligned}$$

The answer is (D).

212. The Soderberg theory states that a fatigue failure will occur if

$$\frac{\sigma_a}{S_e} + \frac{\sigma_m}{S_y} \geq 1$$

and

$$\sigma_m \geq 1$$

It is given that

Yield strength,

$$S_y = 600 \text{ MPa}$$

Alternating stress,

$$\sigma_a = 60 \text{ MPa}$$

Mean stress,

$$\sigma_m = 60 \text{ MPa}$$

Using the data above, the failure condition requires that

$$\frac{60 \text{ MPa}}{S_e} + \frac{60 \text{ MPa}}{600 \text{ MPa}} = 1$$

$$\frac{60 \text{ MPa}}{S_e} = 1 - 0.1 = 0.9$$

Endurance limit,

$$S_e = \frac{60 \text{ MPa}}{0.9} = 66.67 \text{ MPa} \quad (66 \text{ MPa})$$

Rounding the failure stress up to 67 MPa would be unsafe, therefore, it is rounded down to 66 MPa.

The answer is (A).

213. It is given that the bolt is properly preloaded and plates are stiff. Therefore, fatigue is not a serious concern in such tenson-loaded joints. The bolt material is steel, which is relatively ductile. Therefore, the stress concentration is of minor importance. The figure shows a bolted joint in tension, employing a gasket. The moduli of elasticity of steel and bronze are given in the *NCEES Handbook*, Average Mechanical Properties of Typical Engineering Materials table.

step 1: Define load distribution. The assembly consists of a steel bolt and a bronze gasket. The distribution of load between the bolt and the gasket depends on their respective stiffnesses. The stiffness, k, is defined as the load on the system (or an element) that produces a unit deformation. Since the fastener assembly undergoes uniform deformation, the load on the joint assembly is distributed between the bolt and the gasket in the ratio of their stiffnesses. The part of the load carried by the gasket is

$$P_{\text{gasket}} = \frac{k_{\text{gasket}}}{k_{\text{gasket}} + k_{\text{bolt}}} P_{\text{total}}$$

The axial stiffness of an element is defined is given by

$$k = \frac{EA}{L}$$

step 2: For a steel bolt,

Area,

$$A_{\text{bolt}} = \frac{\pi d^2}{4} = \frac{\pi (0.75 \text{ in})^2}{4} = 0.4418 \text{ in}^2$$

$$E_{\text{steel}} = 29 \times 10^6 \text{ psi}$$

Length,

$$L_{\text{bolt}} = 0.75 \text{ in} + 0.25 \text{ in} + 0.75 \text{ in} = 1.75 \text{ in}$$

Therefore,

$$k_{\text{bolt}} = \frac{A_{\text{bolt}} E_{\text{steel}}}{L_{\text{bolt}}} = \frac{(0.4418 \text{ in}^2)\left(29 \times 10^6 \ \frac{\text{lbf}}{\text{in}^2}\right)}{1.75 \text{ in}} = 7.321 \times 10^6 \text{ lbf/in}$$

step 3: Determine the stiffness area of the bronze gasket. See the *NCEES Handbook*, Statics section, for the ring-shaped area of friction surface.

$$A_{\text{gasket}} = \pi(a^2 - b^2) = \pi\left(\left(\frac{2 \text{ in}}{2}\right)^2 - \left(\frac{1 \text{ in}}{2}\right)^2\right) = \frac{3\pi}{4} \text{ in}^2 = 2.356 \text{ in}^2$$

From properties table in the *NCEES Handbook*,

$$E_{\text{bronze}} = 15 \times 10^6 \text{ psi}$$

Use the stiffness formula to determine the stiffness of the gasket.

Length,

$$L_{\text{gasket}} = 0.25 \text{ in}$$

Therefore,

$$k_{gasket} = \frac{A_{gasket}E_{bronze}}{L_{gasket}} = \frac{(2.356\ \text{in}^2)\left(15 \times 10^6\ \frac{\text{lbf}}{\text{in}^2}\right)}{0.25\ \text{in}} = 141.4 \times 10^6\ \text{lbf/in}$$

step 4: Determine the percentage of load taken by the gasket. Assume the total load to be 100 units and use the load distribution formula.

$$P_{gasket} = \left(\frac{k_{gasket}}{k_{gasket} + k_{bolt}}\right)P_{total} = \left(\frac{141.4 \times 10^6\ \frac{\text{lbf}}{\text{in}}}{141.4 \times 10^6\ \frac{\text{lbf}}{\text{in}} + 7.321 \times 10^6\ \frac{\text{lbf}}{\text{in}}}\right)100 = 95.08 \quad (95\%)$$

The answer is (D).

214. See the *NCEES Handbook*, Mechanical Engineering: Equivalent Spring Constant section. The springs are in series. The total elongation of the system is the sum of elongations of each spring.

$$\delta = \delta_1 + \delta_2 = \frac{P}{k_1} + \frac{P}{k_2} = \frac{P}{5\ \frac{\text{kN}}{\text{m}}} + \frac{P}{10\ \frac{\text{kN}}{\text{m}}} = 0.3P\ \text{m/kN}$$

The stiffness of the system, k, is

$$k = \frac{P}{\delta} = \frac{P}{0.3P} = 3.333\ \frac{\text{kN}}{\text{m}} = 3333\ \text{N/m} \quad (3300\ \text{N/m})$$

The answer is (B).

215. Use the *NCEES Handbook*, Statics section. The torque, M, required in lowering a load P using a screw-jack with square thread with a radius is

$$M = Pr\tan(\phi - \alpha)$$

step 1: Simplify the above formula using the trigonometrical identity. See the *NCEES Handbook*, Mathematics section.

$$\tan(\phi - \alpha) = \frac{\tan\phi - \tan\alpha}{1 + \tan\alpha\tan\phi}$$

The friction coefficient equals

$$\tan\phi = 0.12$$

Given

$$P = 500\ \text{kN}$$
$$r = 30\ \text{mm} = 0.03\ \text{m}$$

step 2: Determine the tangent of pitch angle α.

$$\tan\alpha = \frac{p}{2\pi r} = \frac{8\ \text{mm}}{(2\pi)(30\ \text{mm})} = 0.04244$$

step 3: Determine torque.

$$M = Pr\left(\frac{\tan\phi - \tan\alpha}{1 + \tan\alpha\tan\phi}\right) = (500\ \text{kN})(0.03\ \text{m})\left(\frac{0.12 - 0.04244}{1 + (0.04244)(0.12)}\right) = 1.158\ \text{kN·m} \quad (1.16\ \text{kN·m})$$

The answer is 1.16 kN·m.

216. The force of friction is used to transmit power from the engine to a driven shaft. The force of friction is used to start the driven shaft from the rest and to bring it to a proper speed. A single disc or plate with two friction surfaces is shown.

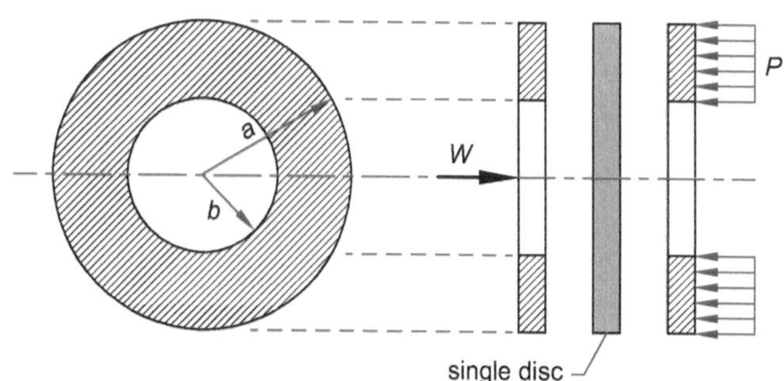

step 1: Determine pressure area. See the *NCEES Handbook*, Statics section, for the ring-shaped area of friction surface.

$$\begin{aligned} A &= \pi(2a^2 - 2b^2) \\ &= \pi\left(\left(\frac{300\ \text{mm}}{2}\right)^2 - \left(\frac{200\ \text{mm}}{2}\right)^2\right) \\ &= 39{,}270\ \text{mm}^2 \end{aligned}$$

step 2: Determine thrust. The mean pressure on the friction is $p = 0.15\ \text{N/mm}^2$. The total force acting on the friction surface is

$$\begin{aligned} W &= pA \\ &= \left(0.15\ \frac{\text{N}}{\text{mm}^2}\right)(39{,}270\ \text{mm}^2) \\ &= 5890\ \text{N} \end{aligned}$$

step 3: Determine applied torque. The torque applied at a ring face is the frictional force on the ring acting tangentially at the mean radius of the ring.

$$\begin{aligned} T_{\text{applied}} &= W \times \text{mean radius} \\ &= (5890\ \text{N})\left(\frac{a+b}{2}\right) \\ &= (5890\ \text{N})\left(\frac{150\ \text{mm} + 100\ \text{mm}}{2}\right)\left(\frac{1\ \text{m}}{1000\ \text{mm}}\right) \\ &= 736.3\ \text{N·m} \end{aligned}$$

The transmitted torque is limited by the friction force the disc face can develop.

$$\begin{aligned} T_{\text{transmitted}} &= T_{\text{applied}} \times \text{friction coefficient} \\ &= (736.3\ \text{N·m})(0.3) \\ &= 220.9\ \text{N·m} \end{aligned}$$

step 4: Determine the power transmitted by the clutch. Transmitted power depends on the shaft speed and the number of disc faces participating.

$$\begin{aligned} \text{Clutch speed} &= 2000\ \text{rpm} \\ &= \left(2000\ \frac{\text{rev}}{\text{min}}\right)\left(\frac{1\ \text{min}}{60\ \text{s}}\right)\left(2\pi\ \frac{\text{rad}}{\text{rev}}\right) \\ &= 209.4\ \text{rad/s} \end{aligned}$$

In this case two disc faces are participating. Therefore,

$$\begin{aligned} \text{Clutch power} &= \text{Torque} \times \text{speed} \times \text{No. of surfaces} \\ &= (220.9\ \text{N·m})\left(209.4\ \frac{\text{rad}}{\text{s}}\right)(2)\left(\frac{1\ \text{kW}}{1000\ \text{W}}\right) \\ &= 92.51\ \text{kW} \quad (93\ \text{kW}) \end{aligned}$$

The answer is (B).

217. A riveted joint can fail through the following modes: shear, rupture, or bearing. The allowable load is the least of the three. See the *NCEES Handbook*, Mechanical Engineering section.

$$\begin{aligned} \text{Rivet } x\text{-section area} &= \left(\frac{\pi}{4}\right)(22\ \text{mm})^2 \\ &= 380.1\ \text{mm}^2 \end{aligned}$$

step 1: Failure of rivets by shear. The rivets are in double shear.

$$\begin{aligned} F_{\text{sh}} &= \left(\begin{array}{l} \text{allowable shear stress} \\ \quad \times x\text{-section area/rivet} \\ \quad \times \text{ No. of rivets/row} \\ \quad \times \text{ No. of shear failure surfaces} \end{array}\right) \\ &= \tau A_{\text{rivet}} n_{\text{rivets}}\ n_{\text{shear}} \\ &= \left(100\ \frac{\text{N}}{\text{mm}^2}\right)(380.1\ \text{mm}^2)(3)(2) \\ &= 228{,}060\ \text{N} \\ &= 228\ \text{kN} \end{aligned}$$

step 2: By observation, the cross-sectional area of the center plate is less than the sum of the areas of the outer plates. Therefore, check the capacity of the center plate. It requires calculating net plate width.

$$\begin{aligned} \text{Net plate width} &= \text{plate width} \\ &\quad -(\text{hole dia})(\text{No. of holes}) \\ &= 300\ \text{mm} - (22\ \text{mm})(3) \\ &= 234\ \text{mm} \end{aligned}$$

$$\begin{aligned} F_t &= \left(\begin{array}{l} \text{allowable tearing stress} \\ \quad \times \text{plate thickness} \times \text{net plate width} \end{array}\right) \\ &= \sigma_t t_{\text{plate}} w_{\text{plate}} \\ &= \left(150\ \frac{\text{N}}{\text{mm}^2}\right)(20\ \text{mm})(234\ \text{mm}) \\ &= 702{,}000\ \text{N} \\ &= 702\ \text{kN} \end{aligned}$$

step 3: Failure by rivet crushing. By observation, check the middle plate since it is stressed twice that of the outer plates.

$$F_b = \begin{pmatrix} \text{allowable bearing stress} \\ \times\text{plate thickness} \\ \times\text{projected plate width} \\ \times\text{No. of rivets/row} \end{pmatrix}$$
$$= \sigma_{\text{br}} t_{\text{plate}} w_{\text{plate}} N_{\text{rivets}}$$
$$= \left(300\ \frac{\text{N}}{\text{mm}^2}\right)(20\ \text{mm})(22\ \text{mm})(3)$$
$$= 396{,}000\ \text{N}$$
$$= 396\ \text{kN}$$

The allowable tensile force the joint can resist is the least of the three forces causing the above failure modes.

$$T_{\text{allowable}} = 228\ \text{kN}$$

The answer is (B).

218. Reliability is defined as the probability that a product will perform successfully under specified operating conditions for a prescribed period. Reliability, like probability, varies between 0 and 1. The problem concerns 20 independent units connected in series. See the *NCEES Handbook*, Industrial and Systems Engineering: Reliability section. The reliability is defined as

$$R(p_1,\ p_2,\ p_3, \cdots,\ p_{20}) = \frac{\sum_{i=1}^{20} p_i}{20}$$

step 1: The data show that out of 20 units tested, 15 units performed at or above the warranty period of 10,000 hr. However, 5 units failed prematurely. A unit is considered reliable if it reaches its full warranty life. A unit is considered unreliable if fails before the warranty life is over. In the reliability analysis, the operating time up to the warranty time is considered as shown. Determine the weighted average of the MTTF.

n	θ (hr)	p_i	$n \times p_i$
15	10,000	10,000 hr/10,000 hr = 1.0	15.0
1	9000	9000 hr/10,000 hr = 0.9	0.9
1	8000	8000 hr/10,000 hr = 0.8	0.8
1	7000	7000 hr/10,000 hr = 0 .7	0.7
1	6000	6000 hr/10,000 hr= 0.6	0.6
1	5000	5000 hr/10,000 hr = 0.5	0.5
$\Sigma =$	18.5		

step 2: Using the formula, determine reliability.

$$R = \frac{18.5}{20}$$
$$= 0.925$$
$$= 92.5\%$$

The answer is 92.5%.

219. See the *NCEES Handbook*, Mechanical Engineering section. The ASME Y-14.5 (2009) on dimensioning and tolerancing, §1.4 applies.

§1.4(h) states that "wires, cables, sheets, rods, and other materials manufactured to gage or code numbers shall be specified by linear dimensions indicating the diameter or thickness. Gage and code numbers may be shown in parenthesis." The rationale is that the gage number is an ambiguous property of a sheet metal as may vary with the metal and the other factors. Therefore, statement A is incorrect.

§1.4(e) states that the drawings should specify a part without specifying manufacturing methods. In other words, the means and methods of manufacturing should be left to the manufacturer. Therefore, statement B is incorrect.

§1.4 states that states that dimensioning and tolerancing shall clearly define engineering intent. However, §1.4 (i) states that a "90° angle applies where center and lines depicting features are shown in 2D orthographic drawing at right angle and no angle is specified." §1.4(j) further illustrates it. In other words, for squares and rectangular shapes, it is implied that their sides have right angles and the drawing need not state the 90° angle. Statement C is incorrect.

According to §1.4(c), each necessary dimension of an end product shall be shown. No more dimensions than those necessary for complete definition shall be given. In other words, mention all necessary dimensions once. Multiple mentioning of a dimension sought should be avoided. Therefore, statement D is correct.

The answer is (D).

220. See the *NCEES Handbook*, Mechanical Engineering section. Three definitions given in the *NCEES Handbook* are relevant: least material condition (LMC), maximum material condition (MMC), and virtual condition. Based on the dimensioning and tolerancing specified in the problem, the hole diameter can vary from 0.990 to 1.010.

Since the cylinder has a tolerance of ± 0.010, the cylinder's limiting dimensions are

$$\begin{aligned}\text{Maximum diameter of the cylinder} &= 0.990 - 0.010 \\ &= 0.980 \\ \text{Nominal diameter of the cylinder} &= 0.980 - 0.010 \\ &= 0.970\end{aligned}$$

According to ASME Y14.5 standard § 1.4 (a), "each dimension shall have a tolerance except for those dimensions specifically identified as reference, maximum, minimum, or stock." Therefore, add tolerance to the nominal dimension. The nominal size of the cylinder is 0.97 ± 0.010.

The answer is 0.97 ± 0.010.